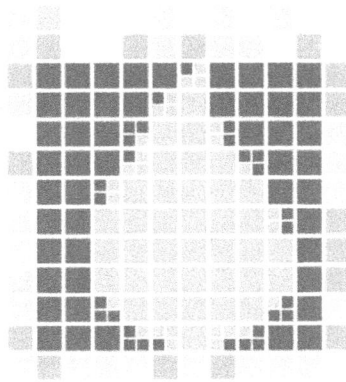

RAFFLES–MAURICK
INTERNATIONAL WATER CONFERENCE

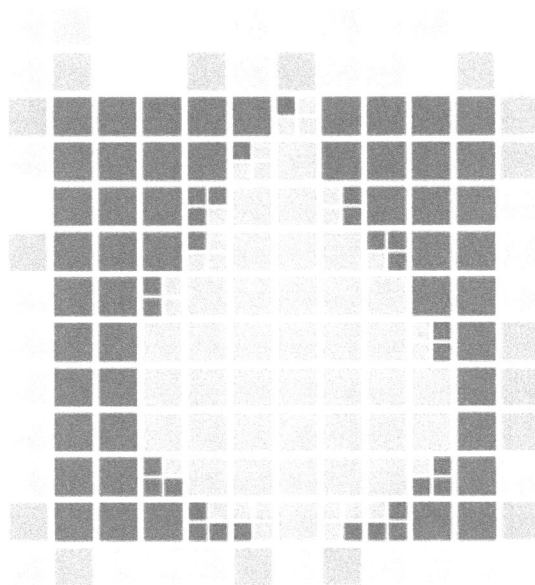

RAFFLES–MAURICK
INTERNATIONAL WATER CONFERENCE

Raffles Institution, Singapore, 9 – 13 June 2014

Editor

Guoxian Tan
Raffles Institution, Singapore

World Scientific

Raffles
Institution

maurick college

Published by

World Scientific Publishing Co. Pte. Ltd.

5 Toh Tuck Link, Singapore 596224

USA office: 27 Warren Street, Suite 401-402, Hackensack, NJ 07601

UK office: 57 Shelton Street, Covent Garden, London WC2H 9HE

and

Raffles Institution
One Raffles Institution Lane
Singapore 575954

and

Maurick College
The Netherlands

British Library Cataloguing-in-Publication Data
A catalogue record for this book is available from the British Library.

RAFFLES–MAURICK INTERNATIONAL WATER CONFERENCE

ISBN 978-981-4632-56-0

CONTENTS

CATEGORY: SCIENCE AND TECHNOLOGY

SUB-CATEGORY: ENGINEERING AND TECHNOLOGY

CATEGORY: COMMUNITY AND LEADERSHIP

SUB-CATEGORY: NATIONAL WATER POLICIES

ACIDIFICATION OF MEDITERRANEAN SEAWATER

Julie Ammendola, Camille Ballet, Lola Bianchi, Eloïse Donnedu

Lycée Honoré d'Estienne d'Orves Nice, France, m.lacour06@laposte.net

Abstract

Study of the link between atmospheric carbon dioxide increase and Mediterranean Sea acidification. Sea water composition, life in sea water (plankton and seashells), dissolution of carbon dioxide in seawater, what the factors which have an influence on that dissolution are and what its impact on marine life on the French Riviera may be.

Connection between carbon dioxide found in the atmosphere and sea acidity. The impact of sea acidity on marine life, especially on zooplankton and seashells.

Keywords

acidification, ocean, seawater, plankton

Introduction

Everyday, oceans absorb about 25 million tons of carbon dioxide CO_2. This sequestration of the CO_2 greenhouse gas, naturally present in the atmosphere, moderates climate changes. However, CO_2 is a polluting acid whose absorption by sea oceans leads to an increase of its acidity: this process is called ocean acidification.

When CO_2 is dissolved in water, it forms carbonic acid H_2CO_3. As human beings emit more and more CO_2 in the atmosphere, oceans have absorbed larger quantities at an increasing rate. This undermines the capacity of the system to adjust to CO_2 natural fluctuations, and drastically changes the oceans' chemistry and makes them more acidic.

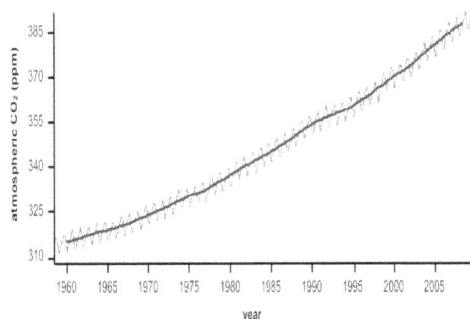

Figure 1 Atmospheric CO_2 fluctuations since 1960 [1]

Since the beginning of industrial era, the global pH has already decreased by 0.1 unit, that is to say an increase of 30% on sea water acidity . In 2000, the average ocean pH fell to 8.1. Scientific researchers consider, in the most pessimistic scenario, that ocean pH should reach 7.6 by 2100.

To enhance our understanding of this ocean acidification phenomenon, we will study those questions:

> **1 What is the composition of the French Riviera sea water ?**
>
> **2 How is sea water acidified?**
>
> **3 What parameters can influence the dissolution of CO_2 in sea water ?**
>
> **4 What are the possible impacts of sea acidification on the environment?**
>
> **5 How do plankton play a fundamental role in gaseous exchanges?**
>
> **6 What are the possible impacts of sea acidification on marine life?**

1 What is the composition of the French Riviera seawater?

Seawater or salt water is water from a sea or an ocean: it is composed of water (**molecules H_2O**) and dissolved salts. Salts are mineral ions: **cations** (positive ions) and **anions** (negative ions).

On average, oceans seawater has a salinity **of about 35 g.L^{-1}**. This means that every kilogram of seawater contains approximately 35 grams of dissolved salts (predominantly **sodium (Na^+) and chloride (Cl^-) ions**).

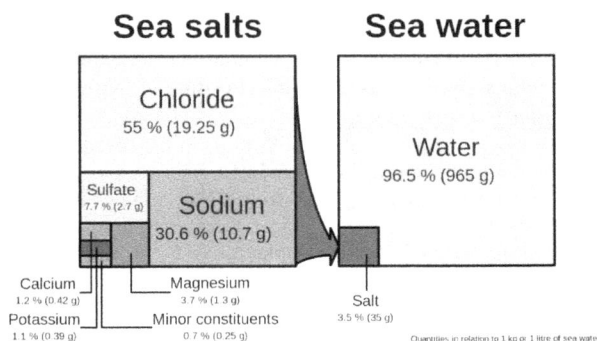

Figure 2 Salts in seawater [2]

1

1.1 How can we measure pH in French Riviera seawater?

First method: use of a pH indicator, the BBT
A pH indicator is a liquid substance which indicates a change of pH in an aqueous solution according to its colour. BBT is:

Blue for pH > 7	Green for pH = 7	Yellow for pH < 7

When we add a few drops of BBT in sea water, the colour is blue: thus sea water pH in Nice is basic.
This method provides only an approximate estimate of the pH.

Second method: use of a pH-meter
A pH meter is an electronic device used for measuring the pH of a liquid. It consists in a special measuring probe (a glass electrode) connected to an electronic meter that measures and displays the pH reading.

The measure of sea water pH with the pH-meter gives the value pH = 8.3
This method provides an accurate value of the pH.

1.2 How can we measure the salinity of French Riviera sea water?

We can suppose that salinity is only due to sodium chloride NaCl salt. Our goal is to determine sea water salinity (in $g.L^{-1}$) collected in Nice Bay by measuring its **electrical conductivity**, considering that the sea water sample has been diluted 100 times.

Water salinity is measured by sending out an electric current between the two electrodes of a conductivity-meter in the sea water sample. The electrical conductivity or EC of a water sample is influenced by the concentration and composition of dissolved salts.

The electrical conductivity of a liquid is proportional to the total amount of dissolved ions for diluted solution.
Method and Procedure:
• We have 4 aqueous solutions S_1 to S_4 in dissolved NaCl: **solvent** = water; **solutes** = Na^+ and Cl^-.
• Mass concentrations C_m in dissolved NaCl are known (see grid below)
• We measure the EC of each solution S_1, S_2, S_3 and S_4 at a temperature of 20 °C
• We measure the EC of sea water S_5 diluted 100 times. We complete the grid below with our results

Solution	S_1	S_2	S_3	S_4	S_5
C_m ($g.L^{-1}$)	0,20	0,30	0,35	0,40	?
EC ($mS.cm^{-1}$)	250	382	442	510	480

• We calculate or draw a graph to deduce the mass concentration of sea water diluted 100 times:

$$C_m(S_5) = \frac{0.40 \times 480}{510} = \frac{0,35 \times 480}{442} = \frac{0.30 \times 480}{382} = \frac{0,20 \times 480}{250}$$

$$\approx 0.38 \ g.L^{-1}$$

> **CONCLUSION 1:** THE SALINITY OF NICE SEA WATER is $C_m = 100 \times C_m(S_5) = 38 \ g.L^{-1}$

2 How is sea water acidified?

We did an experiment to show the acidification after CO_2 dissolution in sea water:

Procedure :
• We pour a few drops of the colour indicator BBT in sea water
• We measure pH sea water with a pH-meter. At the beginning of the experiment pH = 8.3
• CO_2 is blown out with a straw into sea water.

Results:
The colour indicator changed colour (from blue to green/yellow) : from basic, sea water becomes acidic.
The pH meter measures a lower pH : pH= 5.5 after a few minutes.

Therefore, we have demonstrated that CO_2 dissolution in sea water causes the pH to decrease: sea water becomes more acidic.

The dissolution of CO_2 in water causes the solution to become more acidic. So, carbon dioxide can change the pH of water. Indeed, a reaction takes place between carbon dioxide molecules (CO_2) and water molecules (H_2O). Here is how it works :

The CO_2 gas dissolves slightly in water to form a weak acid called carbonic acid, H_2CO_3 **(liquid)**, according to the following reaction:

$$CO_2 \ (g) + H_2O \rightarrow H_2CO_3 \ (l)$$

Then, carbonic acid H_2CO_3 reacts slightly and reversibly in water to form a hydronium cation, H_3O^+, and the **hydrogen carbonate ion, HCO_3^-**; then bicarbonates react to form **carbonate ions CO_3^{2-}**, according to the following reactions:

$$H_2CO_3 + H_2O \leftrightarrows HCO_3^- + H_3O^+$$

$$HCO_3^- + H_2O \leftrightarrows CO_3^{2-} + H_3O^+$$

Changes in carbonate chemistry :

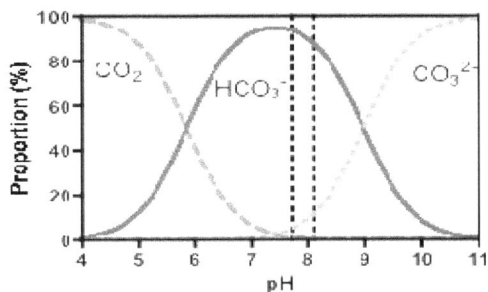

Figure 3 carbonate equilibriums in seawater

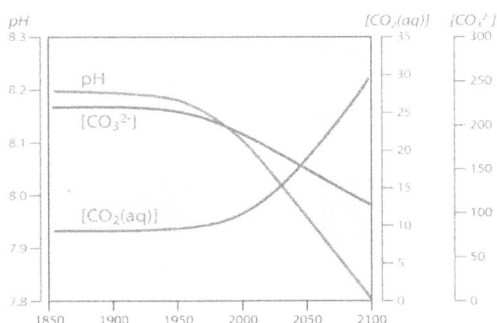

Figure 4 Changes in pH and carbonate concentration while CO_2 increases in seawater

CONCLUSION 2 :
As we can see on the documents, if ocean water pH changes from 8.3 to 7.8, then:
- The quantities of **hydrogen carbonate ions HCO_3^-** will increase; there will still have a lot of them in ocean water.
- The quantities of **carbonate ions CO_3^{2-}** will decrease till they reach a very low level. The problem is that there will not be a lot of them in ocean water.
- The quantities of CO_2 will increase a little.
A CHANGE IN pH OCEAN WATER DRASTICALLY CHANGES THE CARBONATE EQUILIBRIUM

3 Parameters which can influence the dissolution of carbon dioxide in seawater.

We tried to find out potential parameters which might influence the dissolution of carbon dioxide in seawater. We decided to investigate if water **temperature**, and then **salinity**, has an impact on ocean acidification.

3.1 Impact of seawater temperature on CO_2 dissolution

We conducted two experiments:

First experiment : we take a beaker of *hot water* in which we blow a precise amount of CO_2. We measure the pH level at the beginning of the experiment and at the end (12 minutes later).
Indeed we observe that the pH is around 8.12 at 0 min and around 5.97 at 12 minutes.

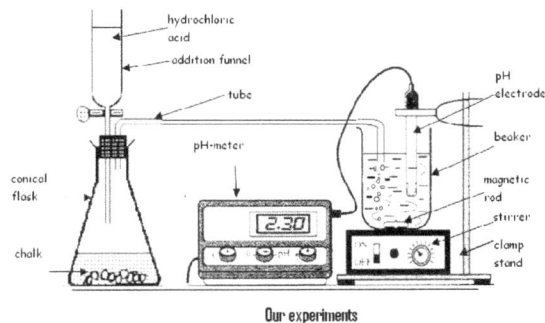

Our experiments

Figure 5 Acidification : lab experiment

Second experiment: we measure the pH value in a beaker with cold water (2°-4° C) in which we also blow exactly the same quantity of CO_2 gas than in the first experiment. The pH is around 8.12 at the beginning too, but at the end it is around 5.17. We can observe that pH level decreases too.

We can observe that the pH level is lower at the end of the second experiment than the first.
So we can conclude that CO_2 dissolves better in cold water than in hot water.

Thus, when the water gets colder, some CO_2 comes through water from the air, on the contrary, when the water gets warmer CO_2 comes through the air from water and increases CO_2 level in the atmosphere.

This phenomenon explains why in cold oceans CO_2 dissolution is enhanced and why ocean acidification is a much more important problem near the poles.

3.2 Impact of seawater salinity on CO_2 dissolution

Then we studied the possible impact of water salinity on CO_2 dissolution:

We took two different beakers of water: In the first one, water has a salinity of 30 g.L^{-1} and in the second beaker, it has a salinity of 40 g.L^{-1}. Then, like in the first experiments, we put exactly the same amount of CO_2 gas and we measure pH level every minute for 12 minutes.

At the end of both experiments, we observe that pH decreases in the two beakers, but the final pH values are the same, around pH = 5.10

We can conclude that salinity has no effect on CO_2 dissolution in ocean water.

CONCLUSION 3

We know that the acidification is the result of CO_2 dissolution in water. Thanks to these two experiments we saw that water temperature has an impact on CO_2 dissolution:
CO_2 DISSOLVES BETTER IN COLD WATER THAN IN HOT WATER.
That is to say ocean acidification is more important near the poles than in the oceans of the equator area
On the other hand, we learnt that water salinity has no impact on CO_2 dissolution.

4 Sea water acidification: impact on the environment

Today oceans absorb about a quarter of human activities CO_2 emissions. It will absorb about 22 million tons of carbon dioxide per day in 2014.

Oceans play the role of a physical and chemical pump: sea currents carry CO_2 across the different oceans; they transport the warm surface water to the North and the cold bottom water to the South. This is called a «treadmill» system. An experiment has shown that if the «treadmill» slows down, the Arctic Ocean will have less capacity to absorb carbon dioxide:

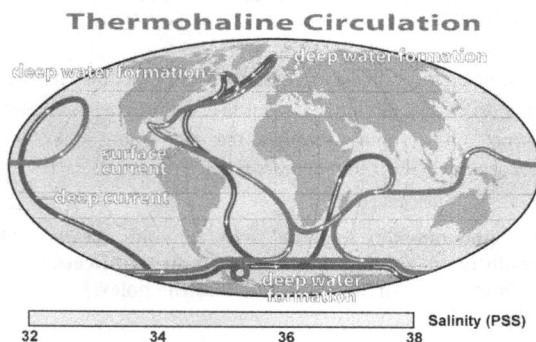

Figure 6 Thermohaline circulation [3]

Some researchers plan that by the end of the century the limit will be reached. When the storage limit of CO_2 in the ocean is reached, the oceans will not be able to continue to absorb the carbon dioxide present in the air. From this moment on, it will stay in the atmosphere. The more CO_2 there is in the air, the more the planet warms up, so this quarter of carbon dioxide previously stored in water, becomes an additional quantity accumulated in the atmosphere. This will have serious consequences on the environment.

CONCLUSION 4
Oceans acidification has a tremendous impact on the environment. If at one point, oceans are not able to absorb CO_2 anymore, then CO_2 will remain in the atmosphere. As CO_2 is the most important greenhouse gas, this will aggravate the Earth global warming.

5 How plankton plays a role in gas exchanges

Oceans play a major role in decreasing the greenhouse effect and therefore global warming thanks to this physical and chemical pump. But it also plays a biological pump thanks to the plankton community: plankton is essential, as it is at the bottom of marine food chain, but also because it exchanges dioxygen and carbon dioxide with the atmosphere. How?

5.1 What is plankton?

Plankton is a multitude of living organisms adrift in the currents. Any living creature carried along by ocean currents is classified as plankton. They range in size from the tiniest virus to siphonophores, the longest animal in the world, and between these two extremes is a world of tiny algae, invertebrates with strange shapes and habits, and countless embryos and larvae.

Our food, our fuel, and the air we breathe originate from plankton. So, plankton plays an important role in human life. Some of them photosynthesize, so they constantly renew the air we breathe. Plankton has also been a great provider of fossil fuels. When planktonic organisms die, they sink onto the sea-bed creating a layer of sediments. Over millions of years, these sediments fossilize, producing our precious oil.

Finally, plankton nourish us: they are the basic part of the food chain in which the large ones eat the small ones. Without plankton, there would be no fish. Plankton is also an important indicator of water health because it is affected by slight changes in the environment, such as temperature or acidity variation…
There are two types of plankton:

Phytoplankton are types of plants: Phytoplankton obtain their energy thanks to the phenomenon of photosynthesis, pulling nutrients from the water surrounding them.

Phytoplankton tend to live near the surface of the water where they can find a lot of sun. They **photosynthesize in the day, releasing O_2, and they are responsible for the production of half the world's O_2**. They also release CO_2 all the time.

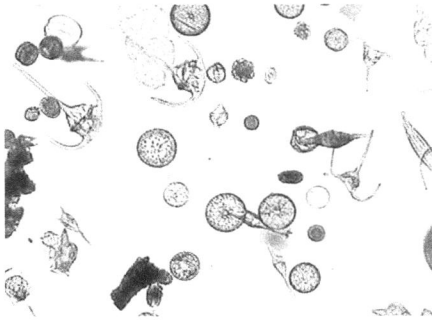

Figure 7 Phytoplankton

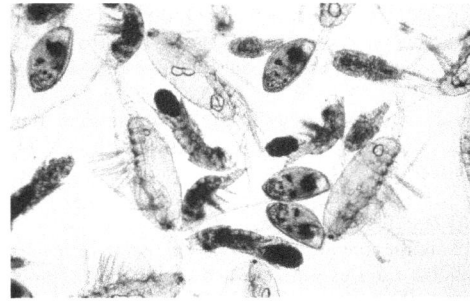

Figure 8 A species of zooplankton

Zooplankton are animals

Zooplankton feed on other plankton, along with bacteria and phytoplankton.

Zooplankton often stay in the deeper parts of the ocean, lakes or ponds where they can find sunlight and then go to the surface at night to feed on other plankton as said above. They use a mechanism called **respiration**: they transform O_2 **into** CO_2 like human beings.

5.2 Plankton and gas exchanges

Thanks to LATIS-BIO probes, we measured the dioxygen and carbon dioxide exchanges between phytoplankton and sea water.

Procedure :

• We put phytoplankton in the LATIS-BIO vessel

• We measure O_2 and CO_2 quantities in the water when phytoplankton are in the light and in the dark.

• We obtain the following results:

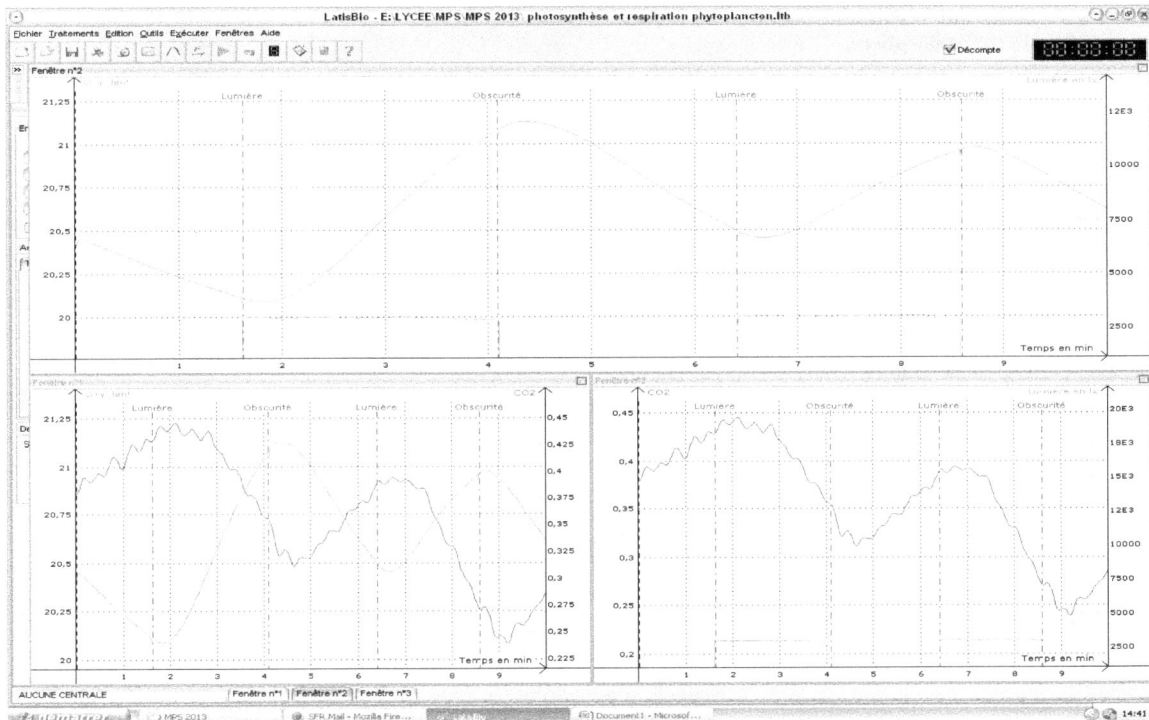

Figure 9 Photosynthesis and respiration results

In the light : O_2 increases, and CO_2 decreases. That means that phytoplankton release dioxygen O_2 and absorb CO_2. In the light, phytoplankton use CO_2 during the photosynthesis process.

5

In the dark : CO_2 increases, and O_2 decreases. That means that phytoplankton releases CO_2 and absorbs O_2. In the light, phytoplankton uses O_2 during the respiration process.

The analysis of both graphs (fig 9) shows that plankton releases more O_2 via photosynthesis than CO_2 via respiration.

A/ PHOTOSYNTHESIS:

Phytoplankton converts carbon dioxide and water into food compounds, such as glucose, and oxygen. This process is called photosynthesis and it needs light. The reaction of photosynthesis is as follows:

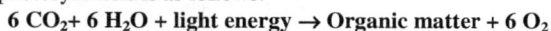

6 CO_2 + 6 H_2O + light energy → Organic matter + 6 O_2

B/ RESPIRATION:

Phytoplankton, in turn, converts food compounds by combining it with dioxygen to release energy for growth and other life activities. This is the respiration process, the reverse of photosynthesis. The respiration reaction is as follows:

Organic matter + 6 O_2 → 6 CO_2 + 6 H_2O

CONCLUSION 5
Photosynthesis has the upper hand on respiration. Plankton absorbs a lot of carbon dioxide CO_2 and rejects it, but they absorb more than they reject. Above all, plankton releases dioxygen O_2.
So the Earth needs our ocean and plankton to reduce CO_2 quantities in the atmosphere.

6 Sea water acidification impact on marine life

Lots of scientific researches confirm the concerns about ocean acidification impacts on certain marine organisms, such as corals and molluscs.

As a matter of fact, we know that acidity dissolves calcium carbonate $CaCO_3$, thus acidity could dissolve marine organisms shells made of calcium carbonate: corals, plankton, calcareous larvae...

Marine living organisms which have a calcareous skeleton need carbonate ions CO_3^{2-} to produce calcium carbonate $CaCO_3$. In part II, we understood that a lower pH leads to less carbonate ions CO_3^{2-} in ocean water, this decrease in carbonate ions could lead to a lower rate of shells formation.

To check these assertions, we did two kind of experiments:

6.1 Experiment with oyster, mussel and shrimp shells

Procedure :
• We pour a few drops of hydrochloric acid on an oyster,

mussel and shrimp shells. The hydrochloric acid has a pH = 5

Results:

• We observe gas bubbles formation on the oyster and mussel shells (inside and outside the shells). This gas is carbon dioxide CO_2 gas.

• The oyster and mussel shells dissolve quickly in acid.

• We observe that there is no gas formation with the shrimp shells.

We did some researches on the internet and found out that shrimp shells are made of chitine which doesn't react to acid, and which is not dissolved in acid, unlike calcium carbonate $CaCO_3$, the main constituent of seashells.

Conclusion: acid dissolves calcareous shells of marine species. In our experiment, the pH of the acid we used was 5.5... ocean water will never reach this pH, except in some areas near submarine volcanoes!!

In part II, a decrease of pH leads to a decrease of carbonates ions CO_3^{2-}, those ions being essential to form the calcareous shells of sea shells... How might the decrease of these ions affect the growth of marine species skeleton?

6.2 Experiment with an urchin larvae*

As it was very difficult to obtain urchin larvae in December, we proceeded to a virtual experiment on the website i2i.stanford.edu

This experiment consists in studying an urchin larva skeleton growth in sea water with different pH:

3 larvae will grow in sea water with pH = 8.1: the present average pH oceans

3 larvae will grow in sea water with pH = 7.7: the estimate pH oceans in 2100

After 6 days, the larvae are observed on a microscope, and their calcareous arms are measured.

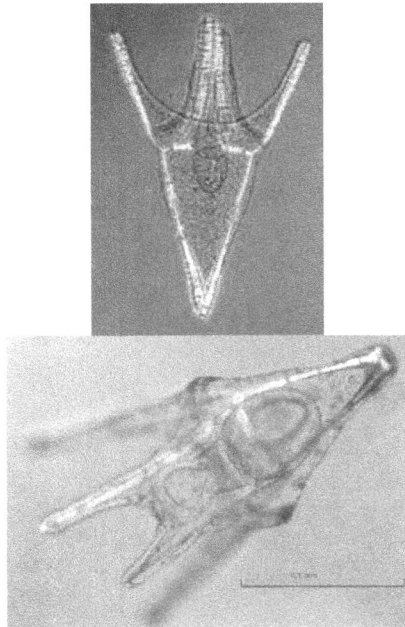

Figure 10 Urchin Larvae

Results after a 6 days ' growth

	Larvae in pH 8.1	Larvae in pH 7.7
Number of larvae	3	3
Average size of arms	533.6 μm	451.2 μm
Standard deviation	± 29.2 μm	± 48.2 μm

We can notice that after 6 days in pH 7.7, the urchin larvae skeleton is smaller than in pH 8.1 because of the acidity. In sea water of pH 7.7 there are less carbonate ions than in sea water of pH 8.1. Therefore urchin larvae have less carbonate to build their skeleton, thus their skeleton growth is slowed down.

Larvae are very vulnerable to predation, so, whatever may slow down or weaken their growth may lead to more larvae being eaten, or even die out…

6.3 Other possible impacts on marine life

Several studies have demonstrated the impact of a lower ocean pH on different marine species:

These results reveal that ocean acidification can have a positive impact on some algae. Indeed, acidification brings more CO_2 in oceans, and CO_2 can promote photosynthesis and improve phytoplankton development.

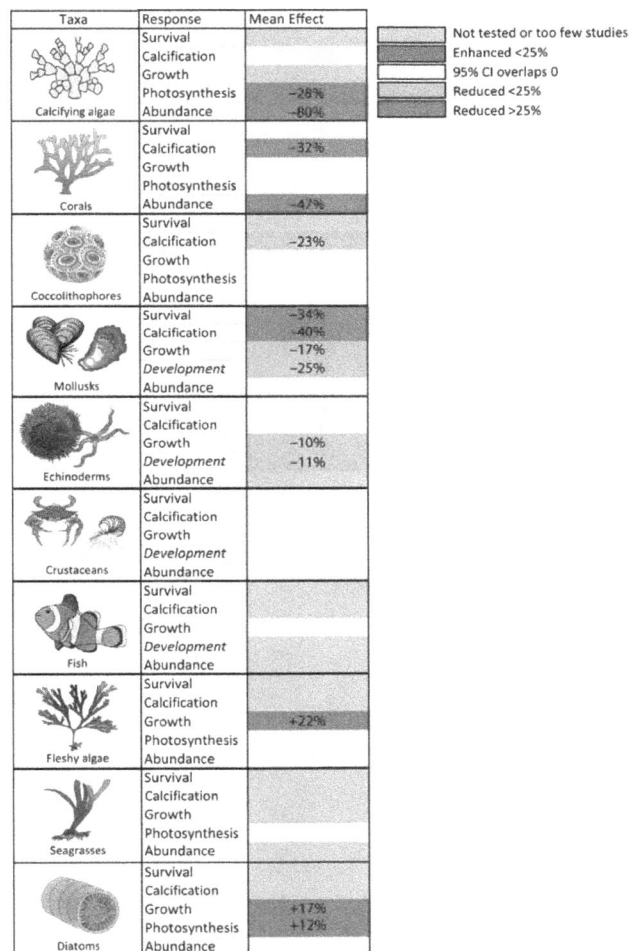

Taxa	Response	Mean Effect
Calcifying algae	Survival	
	Calcification	
	Growth	
	Photosynthesis	−28%
	Abundance	−80%
Corals	Survival	
	Calcification	−32%
	Growth	
	Photosynthesis	
	Abundance	−47%
Coccolithophores	Survival	
	Calcification	−23%
	Growth	
	Photosynthesis	
	Abundance	
Mollusks	Survival	−34%
	Calcification	−40%
	Growth	−17%
	Development	−25%
	Abundance	
Echinoderms	Survival	
	Calcification	
	Growth	−10%
	Development	−11%
	Abundance	
Crustaceans	Survival	
	Calcification	
	Growth	
	Development	
	Abundance	
Fish	Survival	
	Calcification	
	Growth	
	Development	
	Abundance	
Fleshy algae	Survival	
	Calcification	
	Growth	+22%
	Photosynthesis	
	Abundance	
Seagrasses	Survival	
	Calcification	
	Growth	
	Photosynthesis	
	Abundance	
Diatoms	Survival	
	Calcification	
	Growth	+17%
	Photosynthesis	+12%
	Abundance	

Legend:
- Not tested or too few studies
- Enhanced <25%
- 95% CI overlaps 0
- Reduced <25%
- Reduced >25%

Figure11 Responses to lower ocean pH on different marine species [4]

WE CAN CONCLUDE THAT OCEAN ACIDIFICATION CAN HAVE:

Positive effects (on plants like phytoplankton),

Negative effects (on calcareous shelled animals, zooplankton)

And doesn't have any impact on fish and some shellfish like shrimps because their shells are not made of limestone.

7 Conclusion

Experiments suggest that lowering the pH of seawater, and reducing its concentration of carbonate ions CO_3^{2-} will reduce the growth rate of some organisms. Many of these creatures—from plankton and algae to molluscs and corals—form a vital part of the marine ecosystem and may already be threatened by global warming and ocean acidification.

Some of these organisms will adapt. Others will not.

If CO_2 emissions remain unchanged, this rate is expected to increase 170 % by 2100 compared to levels prior to the industrial era. The more ocean acidity will be controlled, the more the ocean's ability to absorb the carbon dioxide emitted into the atmosphere will be reduced, decreasing at the same time the role oceans play in climate change.

TIME TO ACT!

The key to recovery will be to act before the oceans reach a point when they can no longer recover, even if carbon dioxide emissions were cut to pre-industrial levels. This is called the 'tipping point'. When we reach it, the seas could become so acidic that plankton and the shells of winkles, oysters, molluscs, and many other marine animals would start to dissolve.

Sea life could start to dissolve in the Antarctic Ocean by 2030 unless we act now...

Acknowledgements

We would like to thank Mrs Marie Lacour, Mrs Sylvie Ghibaudo, Mrs Claude Romano our Physics, Biology and English teachers and our English teaching assistant Cody Laplante.

Thanks to the C.D.M.M. Centre de Découverte du Monde Marin (Marine world discovery center) in Nice, specially Claire and Mathilde. Thanks to The C.N.R.S. laboratories of Villefranche-sur-mer Oceanologic observatory and specially to Dr. Frédéric Gazeau for his conference.

References

[1,2,3] Wikimedia

[4] Kroeker Kristy J, Kordas Rebecca L, Gattuso Jean-Pierre (2013) *Impacts of ocean acidification on marine organisms: quantifying sensitivities and interaction with warming.* Global Change Biology, jun 2013 ; 19(6):1884-1896

SEA WATER DESALINATION

Leguay Paul, Perrin Margaux, Ulman Melissa, Varé Mayélène

Lycée Honoré d'Estienne D'Orves, France, m.lacour06@laposte.net

Abstract

South-east of France owns large fresh water resources and does not need to desalinate any sea water, nevertheless, we have chosen to investigate some methods in order to understand the technical and ecological stakes linked to water desalination.

Figure 1 View of Eze Village and Mediterranean Sea[1]

Some practical and scientific ways of making fresh water from seawater will be studied. The easiest way being distillation, we will distillate seawater with a solar still and measure the efficiency of this method.

Osmotic pressure to extract pure water from salted water will be studied, as well as reverse osmosis.

The impact of those methods on our environment will be analyzed.

Keywords

Desalination, Distillation, Solar still, Reverse osmosis, Environment

1   Introduction

Earth! Our beautiful blue planet! Most of our planet is covered by oceans that stretch across some two-thirds of its surface, inland, many rivers and lakes offer beautiful landscapes : water seems to be everywhere in most countries.

Unfortunately, very little of that water is available for humans to drink : less than three percent of the planet's water exists as freshwater... and more than two-thirds of this freshwater is frozen in glaciers, in places like the Antarctic and Greenland ice sheets. These resources are mostly inaccessible for human use.

Almost all of the rest of Earth's freshwater resource is groundwater.
Finally, a minuscule percentage of Earth's water is in the form of surface freshwater.

This figure below helps to show that our water resources on Earth is not as abundant as it seems. What is more, fresh water available is really scarce and precious.

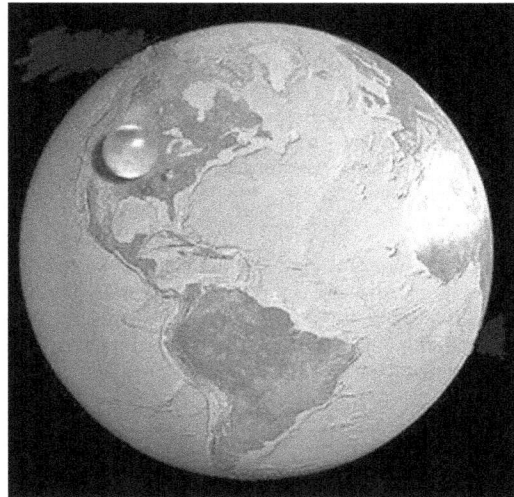

Figure 2 All Earth's water, liquid fresh water, and water in lakes and rivers [2]

Spheres showing:
(1) All water (sphere over western U.S., 860 miles in diameter)
(2) Fresh liquid water in the ground, lakes, swamps, and rivers (sphere over Kentucky, 169.5 miles in diameter), and
(3) Fresh-water lakes and rivers (sphere over Georgia, 34.9 miles in diameter).

In fact, about 97 % of water on Earth is salt water !

As the Earth's population continues to grow and develop, our limited freshwater resources become increasingly scarce.

Therefore, it is crucial for many countries to master sea water desalination techniques.

For a better understanding of desalination issues, we will examine the following points :

2 How water is naturally recycled through the hydrological cycle

3 A first technique used to desalinate water : solar distillation

4 A second method : reverse osmosis

5 Role of activated charcoal in desalination process

6 What are the impacts of desalination on environment

General Conclusion

2   The hydrological cycle

The amount of water on the planet has been roughly constant since the Earth was formed more than four and a half billion years ago.

Water on Earth moves continually through the water cycle of evaporation and transpiration (evapotranspiration), condensation, precipitation, and runoff, usually reaching the sea. Evaporation and transpiration contribute to the precipitation over land.

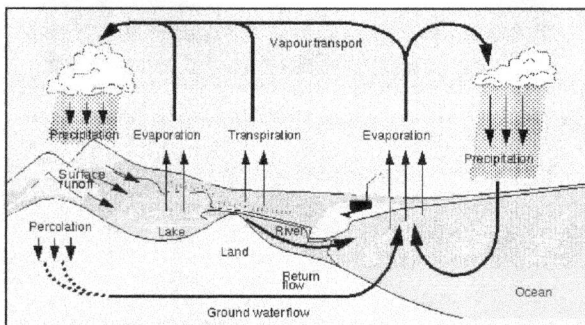

Figure 3 The stages of the Hydrological Cycle [3]

When sea water evaporates, the vapour is pure water, without salt. This is why one of the oldest technique to purify water is solar distillation.

3   Sea water solar distillation

3.1   What is solar distillation?

There are many techniques using solar energy, but all are based on the same principle. The sun heats an object holding unclean or salted water, the water evaporates, and the water droplets are captured. These water droplets are pure, fresh water. Besides seawater, many other sources of

non-potable water can be used including swamp water, urine, and mashed vegetation.

3.2   Experiment: how to distill sea water in Lycée Estienne d'Orves ?

In February, at school in Nice, we set up this experimental device:

Method and Procedure:
• Pour Sea water in a crystallizer
• Place a beaker in the center of the crystallizer
• Cover the crystallizer with thin transparent plastic foil, to achieve an optimal greenhouse effect.
• Place a weight on the plastic foil

Figure 4 Experimental device for solar distillation [4]

• Put the device under sunlight for a few days, surrounded by light reflectors (aluminium sheets)

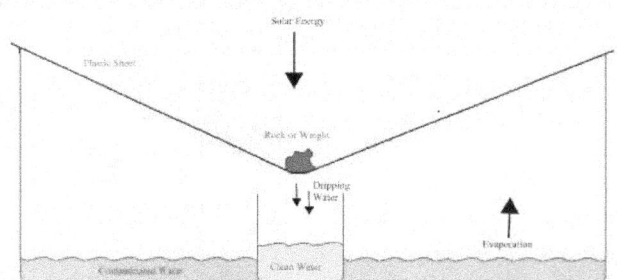

Figure 5 Scheme of the experimental device for solar distillation [5]

Results
After a week, we collect about 1-2 mL of water in the beaker, which is very few.

3.3   How to prove that water in the beaker is non-salted?

We want to check that the water collected in the beaker is fresh water.

We carry out the test :

Experiment →	Silver nitrate — Water from the crystallizer	Silver nitrate — Water from the beaker
Result	White precipitate	No precipitate
Conclusion	Chloride ions are present in sea water	There are no chloride ions in the water from the beaker.

3.4 Quantity of water collected

The water collected after a week is about 1-2 milliliters. The efficiency of our solar distillation is very low. This can be explained by the following facts:

- In february – march, it is winter in Nice: sun-rays are not powerful and air temperature is low (15-17 Celsius degrees). Thus the solar energy transferred to water is not sufficient to ensure a high efficiency.
- The evaporation surface of our device is small. We used crystallizers with a diameter of 40 cm. Thus the surface evaporation is:

$$S = \pi \times R^2 = \pi \times (0.20)^2 = 0.13 \text{ m}^2$$

The amount r of water collected in a week per square meter is:

$$r = \frac{2 \text{ mL}}{0.13} = 15 \text{ mL.m}^{-2} = 1.5.10^{-2} \text{ L.m}^{-2}$$

This yield is really low. Good solar distillers can have a yield of 5-10 L.day⁻¹.m⁻² according to the country and the season.

To improve our device we could:

- Increase the evaporation surface by using larger containers.
- Use better deflectors to focus the sun's rays on the container.
- Use black containers to absorb more solar energy.

3.5 About the efficiency of this method

To be efficient, this technique must be used with optimum conditions of solar light, temperature and air humidity.

That is why this technique cannot be used in any country in the world, but only in hot regions. It is also used for human freshwater consumption on off-shore oil platforms or ships.

4 Reverse osmosis technique

Sea water desalination plants use reverse osmosis technique. We investigate in order to understand this technique.

First of all, we will be studying osmosis, then reverse osmosis.

4.1 What is osmosis?

Definition :
Water from a solution that is less concentrated will have a natural tendency to migrate to a solution with a higher concentration.

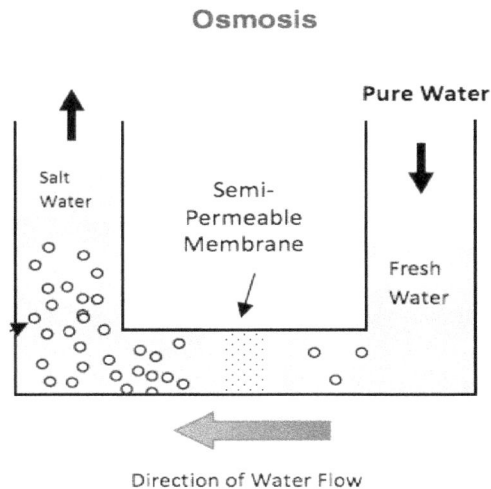

Osmosis

Figure 6 Osmosis principle design [6]

If we had a container full of water with a low salt concentration and another container full of water with a high salt concentration and they were separated by a **semi-permeable membrane**, then the water with the lower salt concentration would begin to migrate towards the water container with the higher salt concentration.

A semi-permeable membrane is a membrane that will allow some ions or molecules to pass but not others.

When a cell membrane is said to be selectively permeable, it means that the cell membrane controls what substances pass in and out through the membrane. This characteristic of cell membranes plays a great role in passive transport. Passive transport is the movement of substances across the cell membrane without any input of energy by the cell.

4.2 Osmosis experiment: mint syrup and sea water

Imagine… You are on a boat in the middle of the Pacific Ocean, you don't have any fresh water, just a chocolate roll wrapped in **cellophane** and some mint syrup. How could you recover drinkable water?

You could set up the following experiment:

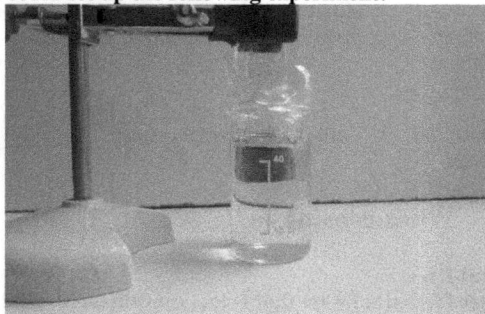

Figure 7 Mint syrup experiment. Step 1[7]

<u>**Procedure:**</u>
• Pour sea water in a beaker
• Wrap the extremity of a glass cylinder with the cellophane foil, held in place with a rubber band.
• Dip the cylinder in the beaker till the rubber band level
• Pour mint syrup in the glass cylinder up to the elastic level
• Wait for 24 hours

The day after, we can observe that the liquid level has increased.

Figure 8 Mint syrup experiment. Step 2 [8]

<u>**Analysis:**</u>
The sea water aqueous solution here is less concentrated than the mint syrup aqueous solution. The cellophane foil plays the role of a semi permeable membrane which allows pure water to pass through. Due to the osmosis phenomenon, water migrates from sea water towards the mint syrup.

<u>**How can we be sure that there is no salt in the mint syrup?**</u>

We can once again use the nitrate silver test for chloride ions.

Figure 9: test with sea water. A white precipitate is formed due to chloride ions.
Figure 10: test with syrup solution. No precipitate formed. There are no chloride ions in the syrup.
Only fresh water passes through the cellophane membrane.

Figure 9 Silver nitrate test in sea water [9]

Figure 10 Silver nitrate test in the syrup solution [10]

4.3 What is reverse osmosis?

Reverse Osmosis is the process of Osmosis in reverse. Whereas Osmosis occurs naturally without energy required, to reverse the process of osmosis you need to apply energy to the more saline solution. A reverse osmosis membrane is a semi-permeable membrane that allows the passage of water molecules but not the majority of dissolved salts, organics, bacteria.

However, you need to « push » the water through the reverse osmosis membrane by applying pressure that is greater than the naturally occurring osmotic pressure in order to desalinate water in the process, allowing pure water through while holding back a majority of contaminants.

Below is a diagram outlining the process of Reverse Osmosis. When pressure is applied to the concentrated solution, the water molecules are forced through the semi-permeable membrane and the contaminants are not allowed through.

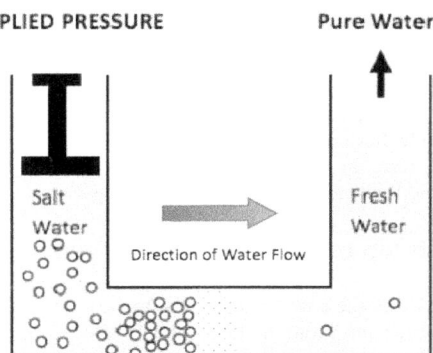

Figure 11 Reverse Osmosis principle [11]

4.4 How reverse osmosis is used in desalination plants

We can consider eight major steps in desalination process:

Step 1: The process starts by extracting water from the ocean using wells located on the shoreline or by using a structure located in the open ocean.

Step 2: The seawater is filtered before passing through the membrane (anthracite, sand, pebbles, gravel...): pretreatment.

Step 3: The second stage of filtration is to the seawater flows through cartridge filters.

Step 4: High pressure pumps increase the pressure of seawater up to 1 000 psig. The aim is to force water from the salt water side through the reverse membranes to the freshwater side.

Step 5: The salt particles from the seawater are rejected from passing through the membranes to the freshwater side. These salt particles remain behind on the concentrated salt water side.

Step 6: The reverse osmosis membranes are put in a fiberglass shell and those membranes are connected end to end. Thanks to the pressure, the flow of freshwater continues and is collected. The concentrated salt stream that is rejected continues to pass across the membrane surface where it is collected separately.

Step 7: The concentrated salt-stream, called brine, will have about 60 % higher salinity. It's then sent back to the ocean in an area significant flow so that the salt particles quickly return to equilibrium with the ocean.

Step 8: Finally, the fresh-water is chemically treated.

As it is impossible for us to achieve reverse osmosis experiments, we decide to work on filtration through cartridge filters. After some research, we found out that lots of filters use activated charcoal and decided to understand its role.

5 Role of activated charcoal in desalination process

5.1 What is activated charcoal?

Figure 12 Activated charcoal [12]

Charcoal is carbon. **Activated charcoal** is charcoal that has been treated with dioxygen to open up millions of tiny pores between the carbon atoms. The use of special manufacturing techniques results in highly porous charcoals that have surface areas of 300-2,000 square metres per gram. These so-called active, or activated, charcoals are widely used to **adsorb** odorous or coloured substances from gases or liquids.

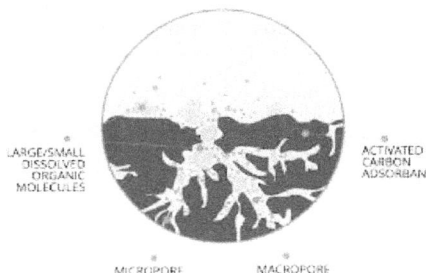

Figure 13 Activated charcoal micropores and macropores [13]

The word **adsorb** is important here. When a material adsorbs something, it attaches to it by chemical attraction. The huge surface area of activated charcoal gives it countless bonding sites. When certain chemicals pass next to the carbon surface, they attach to the surface and are trapped.

5.2 Experiment

Step 1 Preparing an aqueous solution S

• In a 200 mL beaker, add :

- m(NaCl) = 3,5 g of sodium chloride NaCl

- m(CaCl$_2$) = 2,0 g of calcium chloride CaCl$_2$

- several drops of methylene blue (blue solution, the blue colour is due to a big cation)

- several drops of eosin (red solution, the red colour is due to a big anion (see table below))

- Stir in order to dissolve all the constituents and homogenize the solution.

- Filter the solution through activated charcoal

Methylene blue : blue cation	Eosin : red anion

Step 2 Testing for ions before filtration

With the test strips, give the approximate quantities of ions in the prepared solution S :

Chloride ions Cl^- : 0.7 g.L^{-1}

Calcium ions Ca^{2+} : 0.4 g.L^{-1}

Step 3 Testing for ions after filtration

With the test strips, give the approximate quantities of ions in the filtrate and complete the table :

	Colour of the solution	Quantities of chloride ions Cl^-	Quantities of calcium ions Ca^{2+}
Before filtration	purple	0.7 g.L^{-1}	0.4 g.L^{-1}
After filtration	colourless	0.7 g.L^{-1}	0.4 g.L^{-1}

Step 4 Conclusion

- The activated carbon filters :
  - Eosin anions : YES
  - Blue methylene cations : YES
  - Chloride ions Cl^- : NO
  - Calcium ions Ca^{2+}: NO

The activated carbon filters the biggest ions, but lets through smaller ions like chloride and calcium ions.

In a desalination plant, cartridge filters are used but they cannot filter small ions like chloride and sodium ions.

Reverse osmosis process is necessary to complete sea water desalination.

6   Desalination impacts on environment

As mentioned above in paragraph 4.4, desalination treatment produces a large amount of brine which is sent back to the ocean. Therefore, we decide to investigate how brine might have an impact on marine life. We put onion cells, in turn, in an isotonic fluid, hypotonic fluid and hypertonic fluid to observe the cells reactions.

6.1   Onion cells in an isotonic fluid

Isotonic fluid: the concentration of solutes in the solution is **equal** to the concentration of the solutes inside the cell.

Materials :
Onion skin
Slide w. Cover slip
Scalpel
Dropper
Distilled water- tap water
Microscope
Absorbent paper

Procedure:
- Cut a thin section of onion from an outer layer approximately 1cm x 1cm using a scalpel.
- Gently peal the outer layer of the onion skin to get a sample nearly 1 cell thick.
- Create a wet mount slide using tap water.
- Observe the size and shape of the cell and its contents.

- Place several drops of tap water on the slide. Do this by dropping the water on one side of the cover slip while gently absorbing fluid using an absorbent paper on the opposite side.

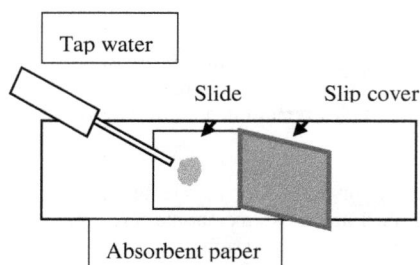

Figure 14 Experiment diagram

Onion plasma membranes are pressed against cell walls.

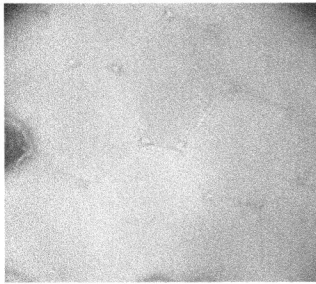

Figure 15 Normal turgidity of onion cells [14]

As the concentration of solutes inside the cell is equal to the concentration solutes outside the cell, water moves equally in both directions and the cell remains equal in size.

The cell has a normal functioning and a normal turgidity.

6.2 Cells in distilled water: hyperturgor

Hypotonic fluid such as distilled water: the solution outside the cell has a lower concentration of than what is inside the cell.

Procedure:
• Do the same experiment
• Place several drops of distilled water on the slide.
• Observe the size and shape of the cell and its contents.

Observation:

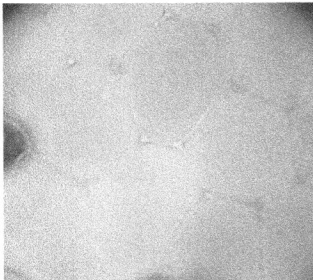

Figure 16 Hyperturgidity of onion cells [15]

Osmosis between the inside and the outside of the cell occurs: water moves from the solution into the cell and the cell swells and can burst open.

This condition is known as hyperturgor.

A slightly salted water causes excessive entry of water into the cells or hyperturgor: inward osmotic flow of water.

6.3 Cells in an highly salted fluid: Plasmolysis

Hypertonic fluid: The solution has a higher concentration of solutes than what is inside the cell.

Procedure:
• Do the same experiment
• Place several drops of 80 g.L^{-1} NaCl solution on the slide.
• Observe the size and shape of the cell and its contents.

Observation:

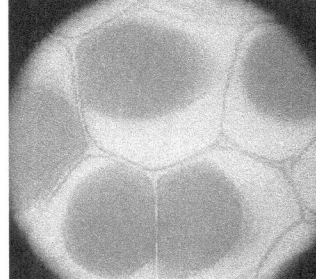

Figure 17 Plasmolysis of onion cells [16]

A very salty water causes excessive water outlet cells or plasmolysis. As a result, water will move from inside cell out into the solution and the cell will shrink in size.

In those conditions, plant cells are not functioning normally, and can be destroyed.

The following diagram synthesises the three situations:

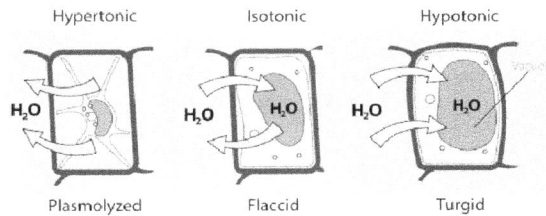

Figure 18 Plasmolyzed, flaccid and turgid cell [17]

6.4. Impact of brine discharge in ocean

As water passes through the reverse osmosis filter membrane, leaving the salt behind, brine is created that is roughly twice as salty as ocean water (about 70-80 g.L^{-1}).

As seen below, high concentrations of salt in discharged water can cause plant and animal cells plasmolysis and eventually lead to the complete collapse of the cell wall.

Numerous studies show that marine plants and organisms die near brine discharge pipes.

It is therefore vital to find solutions, such as diluting this brine with water from the sewage treatment plant before discharging it into the ocean.

Conclusion

Water ! Water is vital for human needs, and, indeed, humans consume more and more of it.

In the world, we use 70 percent of freshwater for irrigation, 22 percent for industry, 8 percent for domestic use.

By 2025, water withdrawals are predicted to increase by 50 percent in developing countries, 18 percent in developed countries.

By 2025, 1 800 million people will be living in countries or regions with absolute water scarcity ; two-thirds of the world population could be under stress conditions caused by water scarcity.

Desalination of sea water is increasingly regarded as an answer, offering the prospect of unlimited water resources, especially in arid countries, despite the cost of producing and the environmental problems that could affect marine life.

Other techniques, like solar distillation, are more accessible to poor countries, but the yield of this technique is rather low.

In any case, we should be aware of the importance of preserving this resource and what we can do to rationalize its use.

If we had more time to study this subject, we would have liked to do further experiments about the impact of brine on marine species, about water filtration in reverse osmosis plants.

Acknowledgements

Thanks to Mrs Marie Lacour, Mrs Sylvie Ghibaudo, Mrs Claude Romano our teachers. Thanks to our English teaching assistant Cody Laplante.

References

[1] [4] [7] [8] [9] [10] our own photos

[2] Credit: Howard Perlman, USGS; globe illustration by Jack Cook, Woods Hole Oceanographic Institution (©); Adam Nieman.

[3] Picture from http://www.euwfd.com/html/hydrological_cycle.html

[5] http://cwanamaker.hubpages.com/hub/How-to-Purify-Water-in-an-Emergency-or-Disaster-Situation

[11] Pictures from http://puretecwater.com/what-is-reverse-osmosis.html

[12] [13] [14] [15] [16] Wikipedia source

[17] http://www.shmoop.com/plant-biology/plant-transportation.html

"The aquifers in Catalonia (Spain): their importance in the water cycle and in the water resource policy in the region"

Alba Abad, Gabriel Antolínez, Jordi Marín, Maria Vericat

Col·legi Mare de Déu del Carme, Spain, meritxell.berruezo@gmail.com

Abstract

The aim of this investigation is to determine whether different natural aquifers respond differently to different types of pollution (brine, nitrates, and fecal contamination). We have chosen three given types of aquifers (limestone, volcanic and sandstone), because they are the most likely to be found in our area. With all this information, we want to develop a ranking of vulnerability of aquifers according to their nature and polluters. Taking into account our proximity to the sea, another risk for our aquifers is the salinization process. We will also determine whether different types of aquifers respond in a different way to this extreme situation. We will also develop a ranking of vulnerability of aquifers according to their nature. through a salinization process.

Keywords

Aquifer, contamination, salinization, environment, education

1. Background information

Often aquifers have not been taken into consideration by governments when the need for water has arisen and this is also the case in Spain. In Catalonia[1], the underground waters are a key factor to water supply for industry and agriculture as well as for drinking water supply. They represent 35% of the resources used, and they exceed 900 hm3/year. Nevertheless, there is a serious problem in the region when it is time to use these resources: there are areas in which the ground-waters are overexploited (mainly coastal areas) and there are areas in which they are not used enough (mainly inland areas). The Catalan government is working towards balancing this situation in an equitable and responsible way.

Acknowledging the importance of a good management of water resources, The European Parliament adopted a Directive in which it was stated that *"...water is essential for human, animal and plant life and is an indispensable resource for the economy"*[2]. This Directive aimed to establish a common framework for the management of surface water and groundwater.

We all need to be aware of the importance of using our natural resources in the right way and water is an essential element to mankind. If our aquifers enjoy a good health and we can avoid pollution and other agents which can be hazardous, we will solve some problems in times of water restrictions, since they are water reserves for periods of drought.

We can also use aquifers to mix this water with the potable water from the water treatment plant. Our water treatment plants can use up to 50% of the river water levels and that amount sometimes is not enough to produce all the drinking water needed in the region. Using water from the aquifers is a good way of getting the amount of water needed. Another advantage is that the water taken from rivers has to go through all the phases of the process, and water from aquifers has only to go through the last stops of the process (chlorination and ozonation), thus making the process a lot cheaper.

By keeping our aquifers in good health, we will have enough water for every use in the region and they will represent a huge economic advantage, especially in the current economic crisis[3] we are going through in the region.

2. Introduction

The aquifers play an important role in the water cycle and they often go unnoticed. There is a human dependence on groundwater. Most land areas on Earth have some form of aquifer underlying them, sometimes at significant depths. These aquifers are rapidly being depleted by the human population.

Fresh-water aquifers, especially those with limited recharge by meteoric water, can be over-exploited and, depending on the local hydrogeology, may draw in non-

---

[1] *Region of Spain which is located on the North-East of the country.*

[2] *European Parliament. 2000. "Water Framework Directive". (WFD)*

[3] *By the end of 2008 Spain entered a recession which is still affecting the country with a high rate of unemployment*

potable water or saltwater intrusion from hydraulically connected aquifers or surface water bodies. This can be a serious problem, especially in coastal areas and other areas where aquifer pumping is excessive. In some areas, the ground water can be contaminated.

Aquifers are critically important in human habitation and agriculture. Deep aquifers in arid areas have long been water sources for irrigation. Many villages and even large cities draw their water supply from wells in aquifers.

Aquifers that provide sustainable fresh groundwater to urban areas and for agricultural irrigation are typically close to the ground surface (within a couple of hundred metres) and have some recharge by fresh water. This recharge comes typically from rivers or meteoric water (precipitation) that percolates into the aquifer through overlying unsaturated materials.

We can find different natures of aquifers in Catalonia. We will be using their characteristics in our research project:

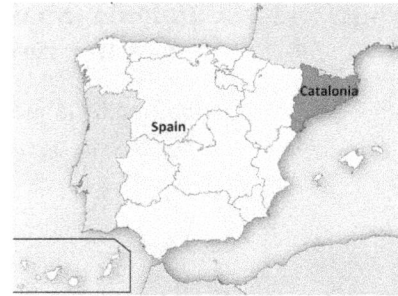

Figure 1: Location of Catalonia (in red)

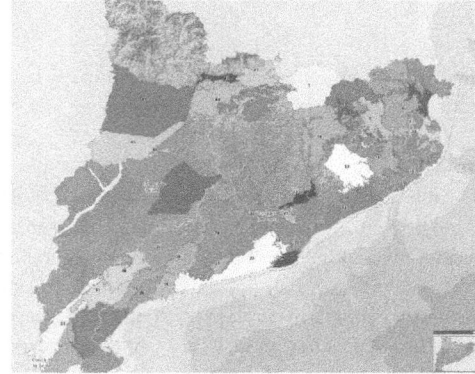

Figure 2: Approx. 1/3 of our aquifers are coastal aquifers

Table 1: Types of aquifers for this project

Types of Flow	Types of Lithology	Name / Location
Mixed aquifers presenting intergranular permeability and/or fissuring	Sandstone materials	Vicfred - Guissona[4] (4 hm³) Baix Gaià[5]- baix Francolí (25hm³)
	Volcanic and sandstone materials	Olot[6]
	Limestone	Tàrrega[7]

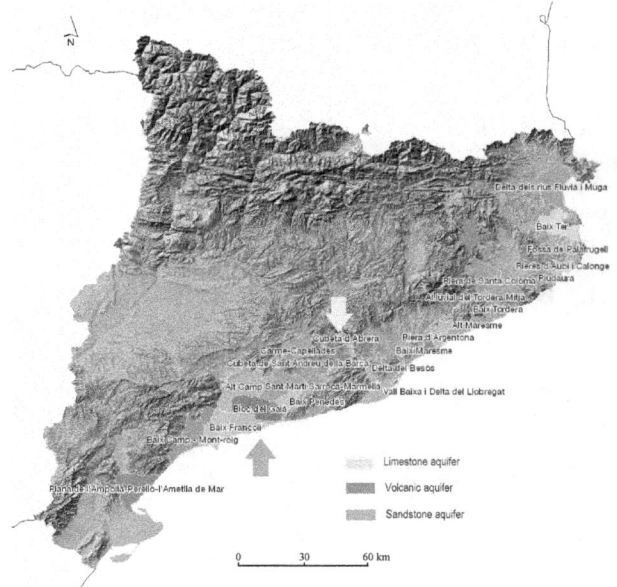

Figure 3: Catalonia aquifers' map

---

[4] Small town of Catalonia whose economy is based upon agriculture and industry.

[5] Small village in the Barcelona province whose economy is based upon tourism.

[6] Large town in the Girona province whose economy is based upon meat industry, metal industry and textiles.

[7] Large town in the Lleida province. Its economy is based upon food and metal industry.

3. Content

3.1. The purpose of the investigation

RESEARCH QUESTION
Do all different natures of aquifers have the same resilience to a contamination process or to an extreme situation (salinization)?

RESEARCH HYPOTHESIS
All types of aquifers may have the same resilience to a contamination process or to an extreme situation (salinization)

3.2. Method of the investigation

Materials

Table 2: List of materials

6 aquariums	pH measurer
10 meters of tubing for each aquarium	Cubes
Fecal contaminator (*Escherichia coli* in culture medium liquid)	Containers for samples
Brine	Tags and markers
Nitrites	Clay
3 suction tubes	Sandstone material
Nitrates, Nitrites and Ammonium test	Volcanic material
Conductivity measurer	Limestone material
Petri dish	Dyes
Medium for bacterial cultivation	Waterproof paint
Stove	Polystyrene
Model trees decoration	Timer
Glue	Fresh water
Distilled water	Micropipettes

* (*See the construction process and the finished aquifers in appendices*)

Methods

First part of the research

In the development of our research, we worked with three aquariums; each aquarium was supervised by one student. Each aquarium reproduces a different type of aquifer present in the Catalan area (see chart above), with its components and characteristics. The three aquifers were filled with three different components:

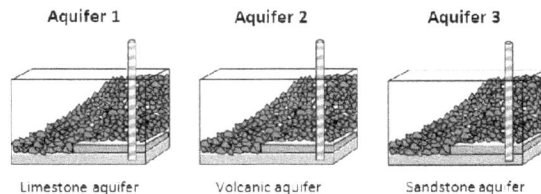

Figure 4: The three types of confined aquifers

In the first part of our research we conducted a pollution process in our aquifer models. We submitted each one of them to different contaminants: brine, nitrites and faecal.

These 3 contaminants were chosen because they are all natural contaminants in our territory (see appendices for some news about aquifer contamination in our region)

The contamination parameter which was analysed in each experimental line was measured before pollution (this sample was our control). The three aquifers were polluted identically in each one of the experiments. After causing contamination we measured the time it took for the aquifer to be filled and emptied in order to return to the initial state (no pollution). After every purge we measured the level of pollution of the water by analysing some water samples.

In order to show how the experiments were developed, here are some links of the processes for each one of the aquifers:

Experiment 1

Brine contamination measured with the conductivity and pH test
Calculate the capacity and recovery time of each and see if there are differences

http://vimeo.com/90240663

http://vimeo.com/90245034

http://vimeo.com/90242309

Experiment II

Nitrates, nitrites and ammonium contamination test
Calculate the capacity and recovery time of each and see if there are differences

http://vimeo.com/90247708

http://vimeo.com/90246822

http://vimeo.com/90257906

Experiment III

Escherichia coli contamination with colony-forming unit (cfu)
Calculate the capacity and recovery time of each and see if there are differences

http://vimeo.com/90325324

http://vimeo.com/90323552

http://vimeo.com/90322723

In all three experiments we monitored the levels of contamination of the outgoing water to determine if there are different responses to this contamination depending on the type of aquifer. The objective was to determine how many times the aquifer had to be refilled until it became pollution-free.

We had three experimental lines and there were as many replications of the experiment as needed in order to have statistical value.

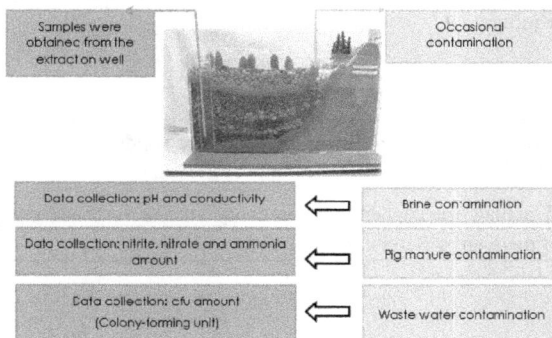

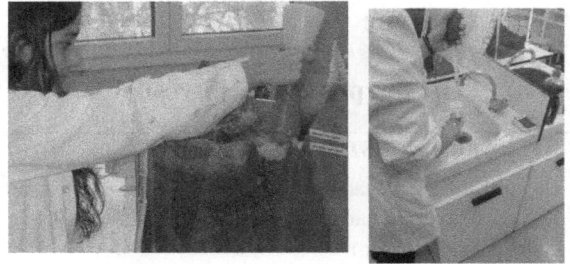

Figure 5: Methodology for contaminating the aquifers

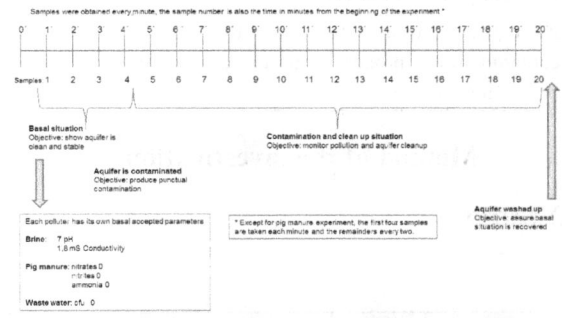

Figure 6: Data collection

Second part of the research

The second part of our investigation was to expose our three aquifers to an extreme situation: salinization.

Once we had recovered our three aquifers from pollution, we exposed them to an extreme salinization. This is a very serious issue and it can be present mainly in those aquifers which are located near the sea if they are overused.

The contamination parameter to be analysed (salinity) was measured before salinizing the water (this was our control sample).

The three aquifers were salinized equally and the aim of the experiment was to see which aquifer could recover first to its pollution-free situation. After causing salinization, we measured the time it took for the aquifer to be filled and emptied in order to return to the initial situation (no salinization). After every purge we measured the level of salinity of the water by analysing some water samples.

We monitored the levels of salinity on the outgoing water to determine if there were different responses depending on the type of aquifer. <u>The objective was to determine how many times the aquifer had to be refilled until it returned to the initial situation.</u>

We had three experimental lines and there were as many replications of the experiment as needed to have statistical value.

In order to show how the experiments were developed, here are some links of the processes for each one of the aquifers:

Experiment IV

| Salinization measured with the conductivity and pH test |
| Observe the differences (if any) in the recovery process and the time they need to do so. |

http://vimeo.com/90317534

http://vimeo.com/90301212

http://vimeo.com/90307663

Salt wedge creation data collection

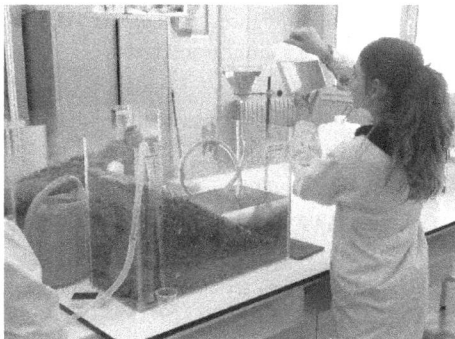

Figure 7: Methodology for creating salt wedge experiments

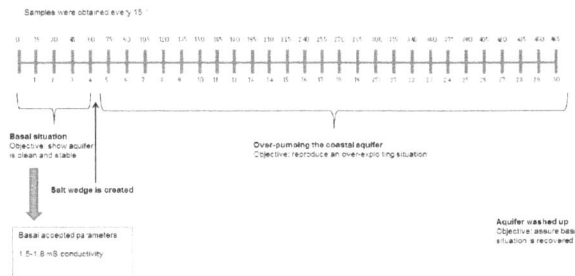

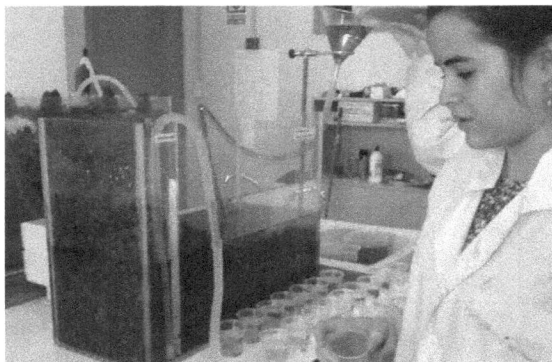

Figure 8: Data collection

Data analysis

Data graphs depict mean ± SEM (standard error of the mean) of the variables studied over time before and after intervention, growth in % from basal to show maximum effect (Emax) and recovery as % of basal in the last sample.

The number of subjects included in each group (normally six for statistical purposes) are shown in each graph as n = x.

Graphs have been produced with GraphPad Prism®. Statistical analysis was performed with Microsoft Excel® and GraphPad Prism®:

1 way ANOVA was performed to compare differences between aquifers before intervention, followed when

statistical significant differences by post-hoc Tukey's t-test to find out which groups differ from others.

Paired t-test was performed to compare 1) basal mean values, 2) Emax induced by intervention and 3) recovery vs basal mean value.

1 way ANOVA was used to compare differences between 1) intervention's Emax, and 2) recovery at last sample, followed when appropriate by post-hoc Tukey's t-test.

Statistical significance was set at the 95% confidence level (two-tailed).

3.3. Results

Confined aquifers results (Contamination)

Brine contamination – effect in pH

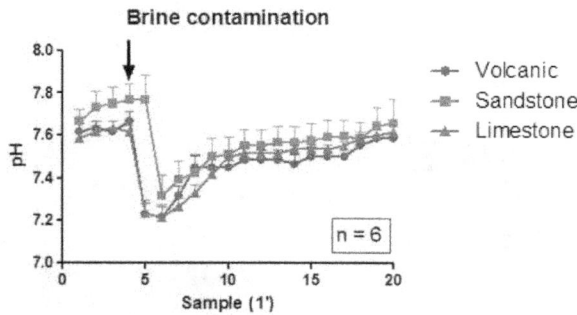

Immediate slight pH decline in the three aquifers after brine contamination (arrow). Steady recovery from sample 6 onwards.

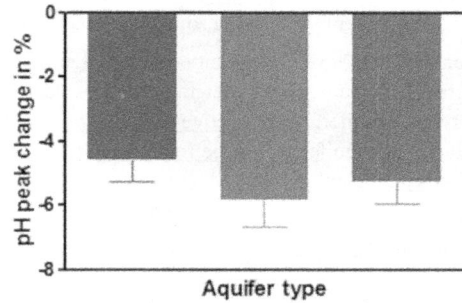

No statistical pH differences between aquifers before contamination.

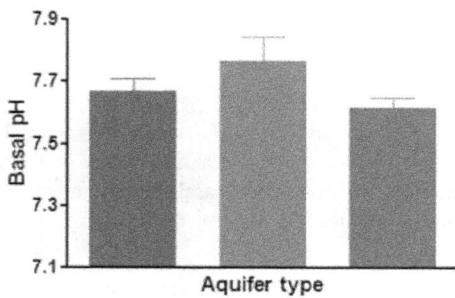

~ 5-6% statistically significant pH peak decline at peak (sample 6). **No statistical effect differences among aquifer types.**

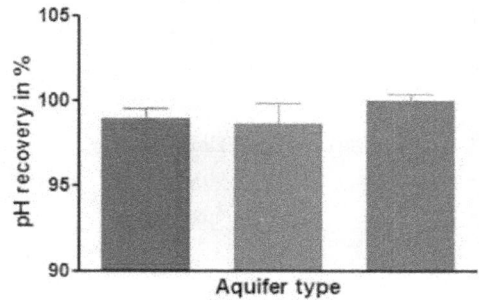

Full aquifer pH recovery at sample 20. No statistical differences vs basal pH values in none of the aquifer types.

Figure 8: Brine contamination effect in pH

Brine contamination – effect in conductivity

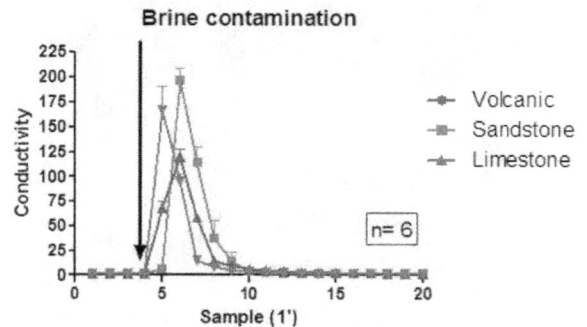

Sharp increase in all aquifers after contamination (arrow). Rapid recovery from sample 6 onwards. Peak conductivity in limestone aquifer reached 1 sample earlier than in other aquifers.

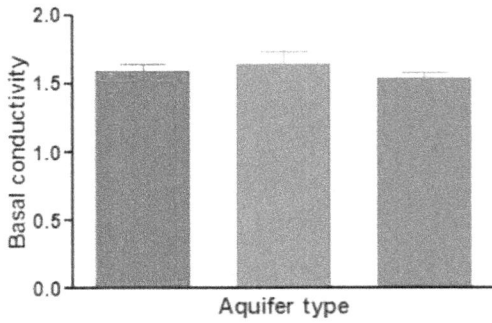

No statistical conductivity differences between aquifers before contamination.

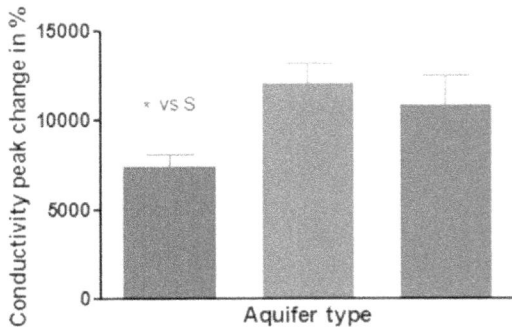

Statistically significant conductivity increase at peak in all aquifers. **Statistical differences in the peak change of conductivity between volcanic and sandstone aquifers**.

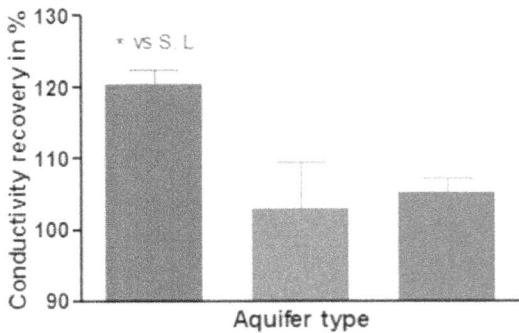

Statistical differences between aquifers concerning conductivity recovery. Full conductivity recovery at sample 20 just in the sandstone aquifer.

Figure 9: Brine contamination effect in conductivity

Pig manure – effect in nitrites, nitrates and ammonium concentration

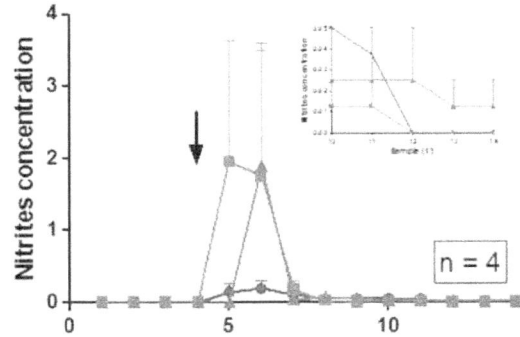

Statistical differences between aquifers concerning conductivity recovery. Full conductivity recovery at sample 20 just in the sandstone aquifer.

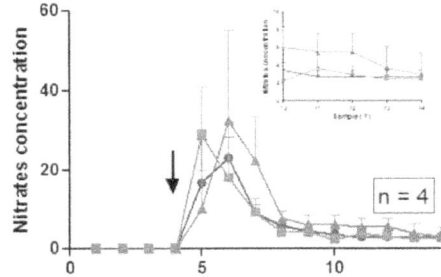

Similar nitrate concentration increase in all aquifers after contamination (no significant differences).

Full wash-out is not seen in any aquifer at sample 14 (no significant differences).

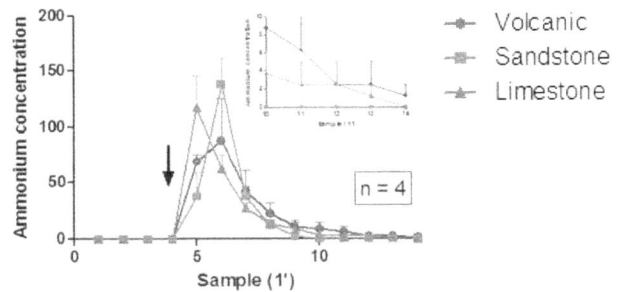

Similar ammonium concentration increase in all aquifers after contamination (no significant differences).

> Full wash-out in sandstone aquifer from sample 10 on. Later recovery in limestone aquifer, whereas not full wash-out is seen in volcanic aquifer.

Figure 10: Pig manure effect in nitrites, nitrates and ammonium concentration

Waste water – effect in cfu

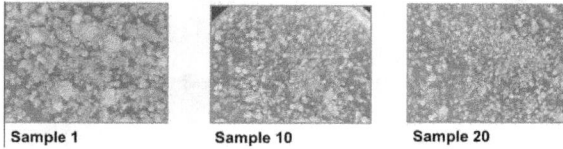

| Sample 1 | Sample 10 | Sample 20 |

Figure 11: cfu obtained in different samples

Not significant differences were appreciate between samples, **even in samples 1 to 4 (before waste water contamination), we founded *E. coli* in all aquifers.**

Aquifers were washed with water with chlorine before the experiment but not enough.

Because of the cfu had different shapes and color it indicates that more species apart from E.coli were in the aquifers.

Coastal aquifers results (Salinization)

Conductivity vs. Salt Concentration

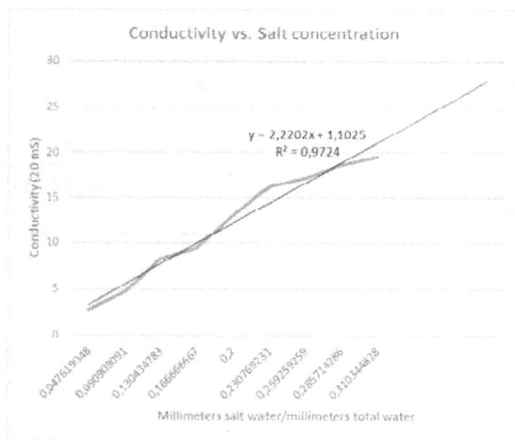

$y = 2,2202x + 1,1025$
$R^2 = 0,9724$

Figure 12: Correlation between conductivity and the salt level in the water

Sea water intrusion – effect in pH

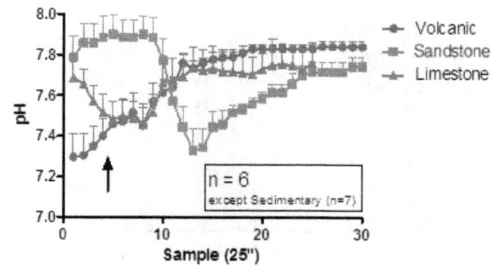

> **pH increase in the volcanic and limestone aquifers whereas reversible pH decrease is seen in the sandstone aquifer** after extreme salinization.

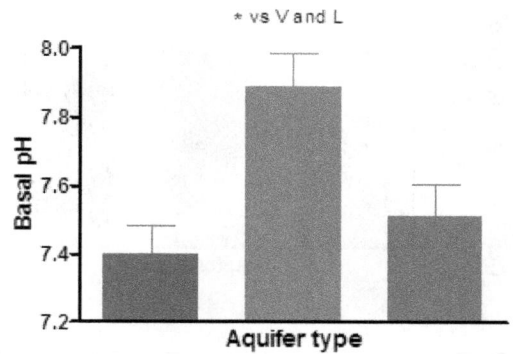

> Statistical pH differences between sandstone aquifers and the others before contamination.

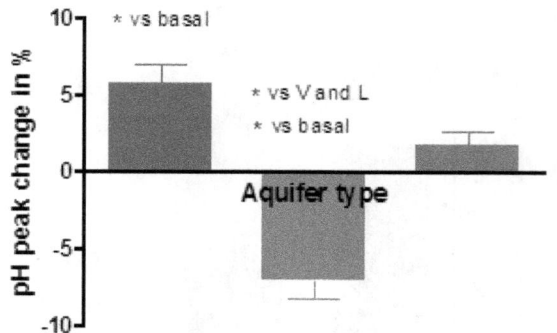

> Statistically significant differences at peak only in volcanic and sandstone aquifers. **The different pH pattern observed in sandstone aquifer is also significant.**

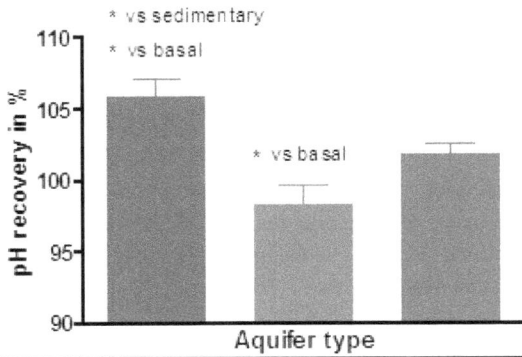

* vs sedimentary
* vs basal

* vs basal

pH recovery in %

Aquifer type

Full pH recovery at sample 30 just in the limestone aquifer.

Figure 13: Sea water intrusion, effect in conductivity

Sea water intrusion – effect in conductivity

Irreversible and statistically significant conductivity increase in all aquifers after extreme salinization.

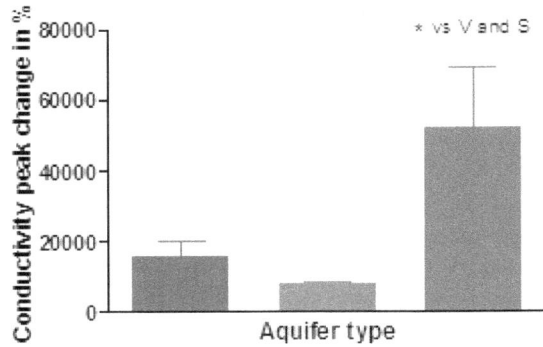

* vs V and L

* vs V and S

Basal conductivity

Aquifer type

Statistical conductivity differences between aquifers before contamination.

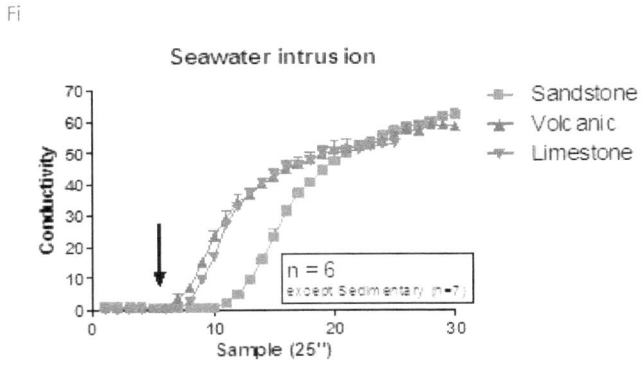

* vs V and S

Conductivity peak change in %

Aquifer type

Statistically significant conductivity increase at peak in all aquifers. Statistical differences in the peak change of conductivity between volcanic and sandstone aquifers.

No analysis for aquifer recovery is shown since maximum salinization is produced at sample 30.

Fi

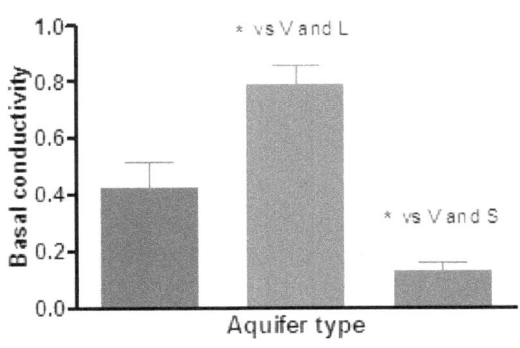

Seawater intrusion

Conductivity

Sample (25'')

n = 6
except Sedimentary (n=7)

- Sandstone
- Volcanic
- Limestone

Figure 14: Seawater intrusion, effect in conductivity

3. Conclusion

Taking into account the typologies of aquifers and the pollutants to which they have been exposed, we are able to conclude that there are indeed varying degrees of sensibilities that affect these aquifers. Different pollutants are more dangerous than others, depending on the nature of the aquifer.

With regard to brine contamination, the most sensitive and most difficult to recover is the volcanic aquifer. Its conductivity is statistically higher to that of the sandstone and limestone aquifers.

As to pig manure contamination, we have observed a significant increment in nitrites, nitrates and ammonium in the volcanic aquifer, which is, in turn, substantially higher than those found in the sandstone and limestone aquifers.

Regarding the different aquifers that have been exposed to waste water contamination, we have not obtained differing results. At first we believed that tap water with chlorine would be sufficient to cleanse the bacteria in the aquifer. However, after the experiments were carried out, we could see that tap water left residual bacterial deposits, as shown by sample 1. Bacteria should not have been found, yet we did.

The fact that we found different types of bacteria would be an ideal starting point for a new line of investigation. A possible title could be "Bacteria and Its Role in Water Cycles"

http://vimeo.com/90521490

Acknowledgments

Mr Xavier Baños (International Operations Manager, Fisher Scientific)

Dr David Brusi (Geology Professor, Universitat de Girona)

Ms Carme Cano (Chemistry teacher, Mare de Déu del Carme)

Dr Josep García (Chemistry PhD, Universitat Politècnica de Catalunya)

Dr Alfons Hervàs (PhD Biochemistry, Institut de Recerca Hospital de Sant Pau)

Ms Sílvia Lope (CDEC chief)

Mr Victor Marín (Student, Mare de Déu del Carme)

Ms Mercè Masip (Geology teacher, Mare de Déu del Carme)

Mr Marc Moreno (Student, Mare de Déu del Carme)

Ms Susanna Serra (Chemistry teacher, Mare de Déu del Carme)

Mr Paul Tompkins (English teacher, Mare de Déu del Carme)

Ms Teresa Garrido (Superficial Water Department Responsible in the ACA)

References

[1] VILARRASA, Victor (2012) *Thermo-hydro-mechanical impacts of carbon dioxide (CO2) injection in deep saline aquifers.* Barcelona. Universitat Politécnica de Catalunya.

[2] SANZ ESCUDÉ, Esteban (2007) *Brackish springs in coastal aquifers and the role of calcite dissolution by mixing waters.* Barcelona. Universitat Politécnica de Catalunya.

[3] BERG, Richard C., KEMPTON, John P. and CARTWRIGHT, Keros (1984) *Potential for contamination of shallow Aquifers in Illinois.* Illinois State Geological Survey.

[4] JIMÉNEZ, Joaquin (2010) *Aquifer recharge from intensively irrigated farmland: several approaches.* Barcelona. Universitat Politécnica de Catalunya. Departament d'Engenyieria del Terreny, Cartogràfica i Geofísica.

[5] *Third report of the United Nations World Water Development Report*

Links of interest

Departament d'Ecologia a Catalunya

http://www.ecologia.cat/eco/index.php?option=com_content&view=article&id=105&Itemid=127

Agencia catalana de l'aigua

http://aca-web.gencat.cat/aca/appmanager/aca/aca?_nfpb=true&_pageLabel=P1228354461208201642682

Global Aquifer Control (GAC) Agencia Catalana de l'Aigua

http://aca-web.gencat.cat/aca/appmanager/aca/aca?_nfpb=true&_pageLabel=P4020022538133216368I695

United States Environmental State Agency

http://www.epa.gov/ogwdw/kids/flash/flash_aquifer.html

Bureau of economic geology

http://www.beg.utexas.edu/education/aquitank/tank01.htm

http://www.eoearth.org/article/Aquifer

USGS

http://water.usgs.gov/ogw/

APPENDICES

APPENDIX I

Where in Catalonia can we find the chosen polluters?

Potash salt mine

Pig farm in Osona

BRINE

Fountains, streams and aquifers in Sallent contaminated by saline waste dump

The sources and the environmental springs recorded in Cogulló chloride levels between 8000 and 100000 milligrams per liter.

Fountains, wells, and streams in the vicinity of the saline waste dump of Cogulló (Sallent) are severely contaminated by brine and saline leach in an area that affects the income of the Llobregat,region. The Superior Court of Justice of Catalonia has cancelled the environmental permission given to the Catalan Government in 2008, authorising Iberpotash Sallent, a big mining company, to extract potash and lay salt residues.

"Everything has been destroyed "says Vendrell Benet, vice president of the neighbours' association of Rampinya, who won the lawsuit. "The salt water coming from the Cogulló sneaks into the earth, destroys the fields and makes everything unusable" says Vendrell.

PIG MANURE

47% of Catalan aquifers located in agricultural production areas are contaminated by pig manure as stated in this article published in one of the main newspapers of Spain:

"It is a serious contamination of the groundwater," answered last month Minister of Territory and Sustainability , Santi Vila, to a parliamentary question from the PSC. Vila was referring to poisoning by nitrates in Catalan aquifers, according to the Catalan Water Agency (ACA), affecting 20 of the 53 groundwater masses, namely 38%. But if we add other pollutants such as trichlorethylene, arsenic and chloride, the masses that do not accomplish quality standards reach 25, namely 47%.

The aquifers which exceeded last year 50 milligrams per liter of nitrate - limit that the World Health Organization recommends not to exceed - are located in the main areas of agricultural production (Lleida , Central Catalonia, Plana de Vic, hollows of Manol and Muga , Baix Penedès and Vallès), and also in the Besòs and Terrassa.

The ACA recognizes that "...agriculture and intensive farming, with a special effect from excess of fertilizes, which causes some of these will end up incorporating into the groundwater". The 87% of nitrates reaching the water come from farming. Much of the manure of livestock, especially pork, is used to fertilize agricultural land. In the last five years the number of pig heads has increased by 20 % to nearly seven billion.

The "Grup de Defensa del Ter" stated that the problem is not the number of pigs, but the management. In recent years, they reported that, "...the intensive farms have multiplied, which are different to small farmers, who look after the environment."

The problem of excess nitrates in water masses is long-standing. From 1998 to 2009, the Government , after consulting the European Union , has declared 12 "vulnerable zones "and that means 33 % of the area of Catalonia. This designation limits the use of nitrate fertilizers to 170 kilograms per hectare. "The big pig farms do not respect it", say from the organization.

The ACA ensures that the main consequence of the excess of nitrates in the water "occurs in water abstraction for public water supply", since 70 % of Catalonia depends on groundwater. The problem is compounded in populations, usually in the inside, that do not have alternative sources, and cannot consume tap water during some periods of the year. The agency spent over seven years almost 40 million euros in "upgrading and replacement of supplies" affected by excess nitrates.

The Government is working to mitigate the effects of swine manure with the Strategic Plan of Land Fertilization and Management of Livestock Manure in Catalonia (2013 - 2016). One of its points refers to the flexibility in the amount of nitrate that can be used in vulnerable areas. "Italy and Belgium already do that, and if it is finally carried out, it will be with the permission of the EU", says the Catalan government.

WASTE WATER

Any city could be exposed to a waste water accident and we should be very aware of the dangers it could bring.

APPENDIX II

<u>**Construction process of the confined aquifers**</u>

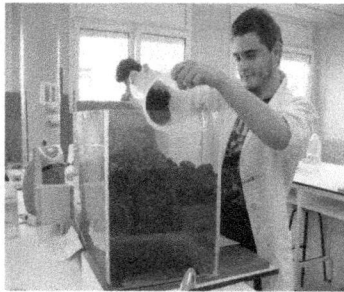

Volcanic confined aquifer

Sandstone confined aquifer

Sedimentary confined aquifer

APPENDIX III

Construction process of coastal aquifers for salinization

Coastal aquifers finished

Sedimentary coastal aquifer

Sandstone coastal aquifer

Volcanic coastal aquifer

Blue Energy

Bente ter Borg, Renee Lunenberg, Dennis van de Sande

Maurick College Vught, Netherlands, Drewes van der Laag: d.vdlaag@maurickcollege.nl

Abstract

Blue Energy is a new way to generate power from the mixing of fresh and salt water. Today, some different methods are applied in order to generate Blue Energy. Within this project, one way will be described and will be part of research-activities. It is called: Reversed Electrodialysis (RED).

In this specific way of generating power, two different kinds of membranes are used in order to create a transport of positive and negative charges. A cationic membrane is used alternately with an anionic membrane. Therefore there will be a charge at outside of the cell containing the membranes. When these charges get in contact with an electrolyte, a redox reaction will be the result. When the two sides of the cell are connected with a metal wire, an electric current will be the result because of the electron transport through the wire.

Obviously, the difference in salt concentration is the initiator of the process. According to this theorem, any cationic and anionic concentration difference can be used in order to generate Blue Energy. In this research the influence of the types of ions and the concentrations of these ions are investigated. The results show that the concentration of the ions and the types of ions influence the generation of energy. A Blue Energy Power Plant (BEPP) nearby the coastal line or near an industrial area (with waste water containing high salt concentrations) may be profitable. A pilot BEPP is installed along the coast line in The Netherlands. The research activities will be executed as a joint venture between Maurick College, Wetsus Leeuwarden and a BEPP.

Keywords

Blue Energy, Reverse Electro Dialysis, waste water recycling, sustainable energy

Introduction

In the past couple of years the emphasis of generating energy in a sustainable way has increased. In 2000 the share of generating energy in a sustainable way was 2,65 %. Nowadays this has increased to 9,08%. [1] One of the projects that contribute to the increase of sustainable energy is Blue Energy. Recently, the REDstack [2] company started building a BEPP at the Afsluitdijk, where energy will be generated using the concept of Blue Energy. In short, Blue Energy is a way to generate sustainable energy by mixing fresh water and salt water. When fresh water and salt water flow in separately and then let them go through some membranes, there will be a charge on both sides of the membranes. As the charges are in contact with an electrolyte, a redox reaction is the result. When both sides are connected with a metal wire, an electron transport will arise and in this way energy will be generated.

The salt water that is used for this project originates from the sea. However, other types of salt water could also be

useful. Salt water could be found in the sea as well in waste water. Salt water from the sea contains sodium and chloride ions. Companies working with chemicals often have waste water that contain ions like sulphate or nitrate. These ions could also be very useful in generating sustainable energy. At the moment, businesses try to filtrate the waste water so that it is not harmful anymore. When the water is firstly used to generate energy, it is already diluted at the end. In this way the waste water is not only recycled, but also made the filtration easier.

Method

Blue energy uses the difference of salt-concentration between two different solutions to create an electric current [3]. Mostly, a solution is used that contains almost no ions, like demineralised water. The other solution often contains a lot of ions and has a high salt-concentration. To make things easy the solutions will be called fresh- and saltwater from now on.

There are different ways to generate energy, using the concept of Blue Energy. The used technique is called: Reverse Electro Dialysis (RED). In contrast to other techniques, RED does not need mechanical parts to generate electricity, except pumps to pump the fresh- and saltwater through compartments and a pump to circulate the electrolyte.

The compartments in Figure 1 are separated by membranes. Those membranes are not all the same.

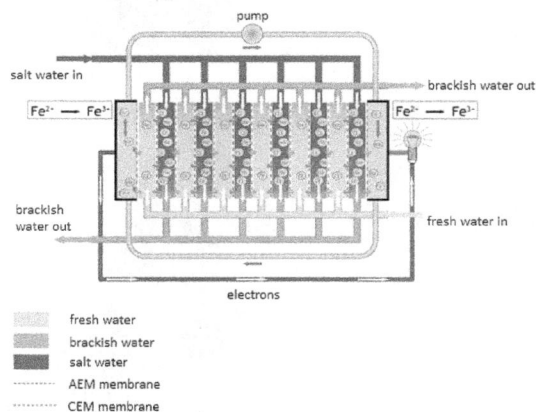

Figure 1: Schematic illustration of RED [4]

There are two different membranes: anion exchange membranes (AEM) and cation exchange membranes (CEM). AEM only allows negative ions through and CEM only the positive ions. Water will not pass through the membranes. In Figure 1 you can see that the positive ions will transport to the right and the negative ions to the left. Eventually a positive current will arise to the right.

Finally, this ionic current needs to be converted into an electric current. This occurs at the electrodes at both sides.

The ionic current causes a redox reaction between Fe2+ and Fe3+ -ions. These reactions create the electric current.

After this process the fresh- and saltwater will be mixed and come out of the compartments as brackish water.

Experiment and results

Some experiments were done at school to see if the concentration of the saltwater influences the voltage. The influence of different anions was checked as well.

Before the measurements were made, the different solutions were made. Sodium sulphate (Na2SO4) and sodium nitrate (NaNO3) were dissolved in demineralised water. Various amounts of these substances were dissolved, to acquire different molarities.

After preparing the solutions, the solutions are inserted in the Blue Energy apparatus the ion containing solution through the saltwater compartments and demineralised water through the freshwater compartments. Instead of a lamp in Figure 1 a voltmeter was used to measure the voltage.

The provisional results are in the tables below:

Substance	Molar mass (g·mol^{-1})	Molarity (mol·L^{-1})	Maximum voltage(V)
Na_2SO_4	142,04	0,500	1,863
Na_2SO_4	142,04	0,250	1,845
Na_2SO_4	142,04	0,100	1,759
Na_2SO_4	142,04	0,050	1,652

Substance	Molar mass (g·mol^{-1})	Molarity (mol·L^{-1})	Maximum voltage(V)
$NaNO_3$	84,99	0,500	2,060
$NaNO_3$	84,99	0,250	1,833
$NaNO_3$	84,99	0,100	1,002
$NaNO_3$	84,99	0,050	1,236

These are the first results within the scope of this research. They are not very reliable, because only one measurement per concentration was made. However, there is one thing that could be concluded from these first results: the higher the concentration of the solutions, the higher the voltage measured.

Other questions are hard to answer right now. More measurements are necessary to see if these results are reliable and to see if the difference between the anions has a significant influence on the voltage.

References

[1] CBS, "Hernieuwbare elektriciteit; bruto en netto productie, import en export," CBS, 25 February 2014. [Online]. Available: http://statline.cbs.nl/. [Accessed 14 may 2014].

[2] REDstack, [Online]. Available: http://www.redstack.nl/. [Accessed 11 march 2014].

[3] Wetsus, "Blue Energy research projects," Wetsus, [Online]. Available: http://www.wetsus.nl/research/research-themes/blue-energy. [Accessed 15 april 2014].

[4] H. Zijlstra, E. Eijkholt, J. Post, J. Veerman, C. v. Oers and J. v. Dalfsen, Blue Energy: zonne-energie uit water, Leeuwarden: Wetsus, 2010.

Canteen wastewater treatment by electrocoagulation

Natnicha Plongmai, Tanya Kurutach and Dr. Usa Jeenjenkit*

Department of Chemistry, Mahidol Wittayanusorn School, Nakhon Pathom, 73170, Thailand
E-mail addresses: usajeen@gmail.com

Abstract

The characteristics of school canteen wastewater were investigated, and it was found that the wastewater has a height amount of oil and grease. Electrocoagulation (EC) is an efficient method for wastewater treatment. Different electrode materials and operational conditions were examined. Aluminium and iron were used as electrodes and they have no significantly different on oil removing efficiency ($p>0.05$) as 87.63% and 80.65% respectively. The parallel electrode arrangement was also found not to be significantly different from series arrangement ($p>0.05$), and their removal capacity of oil and grease are 87.63% and 79.44% respectively. The electrical potentials between 15-30 volts were tested. The experimental results showed that the different electrical potentials can remove oil at 83-87% which is not significantly different ($p>0.05$).

Keywords

Electrocoagulation, Waste water treatment, Oil and Grease

1 Introduction

Canteen wastewater is composed of oil and grease, carbohydrates, starches, proteins, vitamins, pectines and sugars which are responsible for high chemical oxygen demand (COD), biochemical oxygen demand (BOD) and suspended solids (SS). Oil and grease which are characteristic of canteen wastewater are necessary for cooking. They are found more than 1090 mg/L [1] while the standard maximum is 5 mg/L. Furthermore, food scraps passing the litter screen is the main component of canteen waste water. Nowadays, oil removing process uses grease trap that spends a lot of times, so microorganism will decompose them. This is the cause which makes canteen wastewater be rotten and stink.

In recent years, new and novel processes for efficient and adequate treatment have been explored due to strict environmental regulations. Electrocoagulation (EC) process has been attracted a great attention in treating wastewater because of the versatility and the environmental compatibility [2]. This process is characterized by a fast rate of pollutant removal, compact size of the equipment, simplicity in operation, and low capital and operating costs. Moreover, it is particularly more effective in treating wastewaters containing small and light suspended particles, such as oily restaurant wastewater, because of the accompanying electro-flotation effect [1]. Many researches had studied on treating wastewater by EC, and many conditions were investigated in various wastewater.

Therefore, this study has objective to scrutinize the condition appropriate for treating school canteen wastewater in order to solve this problem in our school and other canteens. Moreover, because of their electro-flotation needs lots of metal as electrodes. This study also investigates the effect of electrocoagulation using waste metal electrodes.

2 Content

2.1 Purpose of project

This research has studied the removal capacity of oil and grease from the school canteen's wastewater based on Electrocoagulation using waste metal as electrodes. We aim to determine the suitable condition for the treatment in terms of electrode arrangement, electrode materials and electrical potentials.

2.2 Research Framework

1. Studied conditions

- **Electrode**

  Arrangement
  -Parallel
  -Series

  Types of metal
  -Aluminium
  -Iron

- **Electrical potentials**
  -15V
  - 20V
  - 24V
  - 30V

2. Electrocoagulation using waste metal electrodes: In this study used iron as an electrode from waste metal.

2.3 Method

1. Analyze school canteen wastewater[4]

    1.1 Samples should be visually assessed and checked for pH prior to extraction. If not already done, adjust sample pH to < 2 using HCl or H2SO4.

    1.2 Sample volumes are measured gravimetrically or volumetrically, to at least the nearest 10 mL.

1.3 Samples are sequentially extracted with three aliquots of hexane in a separatory funnel.

1.4 Samples are shaken vigorously for 2 minutes per extraction. The first aliquot of hexane is used to rinse the sample container so that its entire contents are transferred to the extraction vessel. The ratio of solvent to sample should be no less than 1:20, i.e. 50 mL of hexane (per extraction) per 1 L of sample. The solvent extracts are passed through a drying funnel containing anhydrous sodium sulfate and combined together.

1.5 Emulsions frequently occur during the extraction of many oil and grease samples. When encountered, precautions must be taken to ensure that adequate extraction efficiency is obtained.

1.6 For Oil and Grease, the extract is evaporated to dryness at ambient temperature (~20 - 25 °C). Following evaporation, residual water, solvent, and other volatiles are removed by heating in an oven at 50-60 °C for 30 -60 minutes or by continued evaporation at ambient temperature, prior to gravimetric determination of the residue using at least a 4 place balance.

1.7 If the final evaporation step is done at ambient temperature, gravimetric measurements must be done to constant weight (see prescriptive elements).

2 Preparing synthesis water

2.6 Pipet vegetable oil 60.93 mg/L into 3 l of water.

2.7 Pour dishwashing liquid 15 mL. and stir it.

3 Water treatment (EC)

3.6 Prepare synthesis water 500 mL., and six aluminum electrodes sized $3 \times 10 \times 0.12$ cm^3.

3.7 Wire circuit like Figure 1 and use electrical potential 30 voltages. Leave it for 30 minute.

3.8 Measure amount of oil and grease.

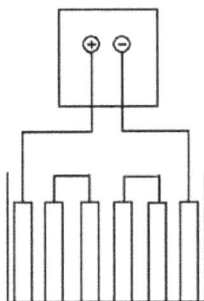

Figure 1: series electrode arrangement

3.9 Repeat 3.1 but change aluminum electrodes to iron electrodes.

3.10 Repeat 3.1 but change series electrode arrangement to parallel electrode arrangement like Figure 2.

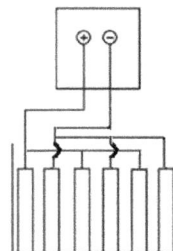

Figure 2: parallel electrode arrangement

3.11 Repeat 3.1 but change electrical potential from 30 voltages to 24, 20 and 15 voltages.

3.12 Repeat 3.1 but change electrode material to waste metal electrodes.

2.4 Results

Table 1: Average quantity and removal capacity of oil and grease in different conditions

Conditons	Average quantity of oil and grease after test (mg/L)	Average removal capacity of oil and grease (%)
Electrode arrangement		
Parallel	7.67	87.63
Series	12.75	79.44
Type of electrodes materials		
Aluminium	7.67	87.63
Iron	12.00	80.65
Electrical potentials		
15V	10.50	83.06
20V	9.67	84.41
24V	10.50	83.06
30V	7.67	87.63

*Quantity of oil and grease in synthesis water before test is 62 mg/L

From the experimental results, we can conclude that all of the studied conditions for school canteen wastewater treatment by EC have no significantly different efficiency.

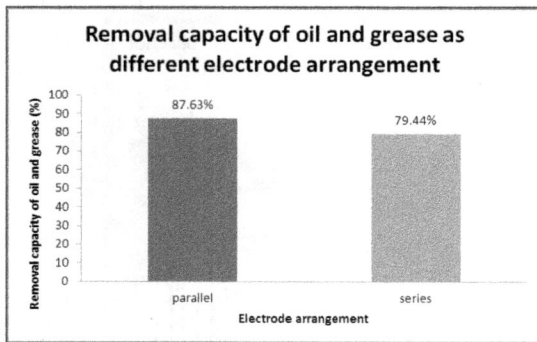

Figure 3: Removal capacity of oil and grease as different electrode arrangement

From figure 3, the parallel arrangement had more average of removal capacity of oil and grease than the series arrangement, but we found that both of these arrangements have no significantly different efficiency from T-test.

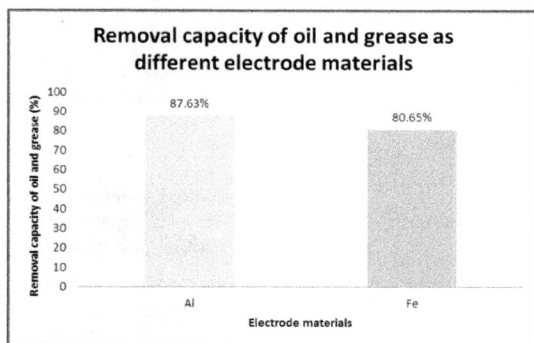

Figure 4: Removal capacity of oil and grease as different electrode materials

From figure 4, Aluminium and Iron could respectively remove 88% and 81% of oil and grease from school canteen wastewater. From T-test can infer that they have no significantly different efficiency.

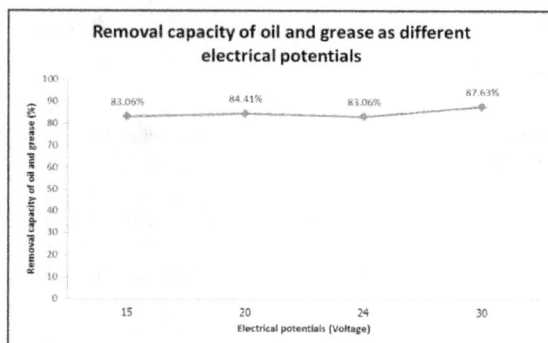

Figure 5: Removal capacity of oil and grease as different electrical potentials

From figure 5, the electrical potential of 30 voltages seem to be the best that can highly remove 88% of oil and grease, but from Anova found that all electrical potentials (15-30V) have no significantly different efficiency.

From studying condition results, we determined to use parallel arrangement, 30V, with waste iron electrode to treat real school canteen wastewater for 30 minutes. We found that this experiment can remove more than 96% of oil and grease and more than 35% of total solid.

3   Conclusion

Our study has shown that different conditions in terms of electrode arrangement, electrode materials, and electrical potentials have no significant difference in their efficiencies. Then, we chose variables suitable for school canteen wastewater treatment by using waste metal electrodes. Those were parallel arrangement, 30 voltages, and waste iron as electrode, and treated the real wastewater for 30 minute. Moreover, we found that waste metal electrodes had a high capability to remove more than 96% of oil and grease from canteen wastewater and more than 35% of total solid.

Therefore, we can conclude that the method of electrocoagulation is an excellent wastewater treatment process. It can be employed efficiently under various conditions.

The future plans
1. Study more conditions and parameters.
2. Developing wastewater treatment by electrocoagulation, and planning about wastewater treatment machine by electrocoagulation.

Acknowledgement

At least we sincerely thank to Mr.Wittawas Phunmunee who helped us prepare electrodes and Mr.Ittipol Sawaddiwongchai who always provided us equipment and advice along this study. Most of all, we deeply thank to our school, Mahidol Wittayanusorn School, which supported and activated us till this study complete.

References

[1]   Chen X., Chen G. and Yue P.L., Separation of pollutants from restaurant wastewater by electrocoagulation. Sep. Purif. Technol., 19 (2000) 65–76.

[2]   Kobya M., Hiz H., Senturk E., Aydiner C., Demirbas E., Treatment of potato chips manufacturing wastewater by electrocoagulation. Des. 190 (2006) 201–211.

[3]   Mollah M.Y.A., Schennach R., Parga J.P. and Cocke D.L., Electrocoagulation (EC)-science and applications. J. Hazard. Mater. B84 (2001) 29–41.

[4]   United States Environmental Protection Agency, Method 1664, Revision A: N-Hexane Extractable Material (HEM; Oil and Grease) and Silica Gel Treated N-Hexane Extractable Material (SGT-HEM; Non-polar Material) by Extraction and Gravimetry. (1999).

Environmentally sustainable solutions to floods in Copenhagen

Ida Maria Hartmann, Nanna Boholt Breitenstein Larsen

Vordingborg Gymnasium & HF, Denmark, lt@vordingborg-gym.dk

Abstract

This study investigates and discusses solutions to floods in Copenhagen, which are environmentally sustainable. Relying on already conducted research, a solution is suggested which will reduce floods as well as securing a sustainable environment. This solution implies construction of a SMART tunnel which substitutes the big road on Åboulevarden, reopening of a former stream which will flow in Åboulevarden towards Skt. Jørgens Sø (a lake), and a water tunnel from Skt. Jørgens Sø to the harbor. The environmental impacts of including the lake in the project, needs further investigation. If these investigations show considerable environmental damage, we suggest an alternative solution, where the Ladegårds Stream continues directly in a water tunnel towards the harbor. The suggested solution will prevent floods, and have various environmental benefits concerning groundwater, surface water, air pollution, and biodiversity.

Keywords

Floods, environment, Copenhagen, tunnel- stream-lake/water tunnel-solution

Introduction

Copenhagen has in recent years experienced heavy rains which have caused floods resulting in severe material damage. It is expected that the frequency and magnitude of these heavy rains will increase in the future, and measures to cope with the heavy rains are therefore necessary.

Denmark is located in the western part of Europe approximately 56 ° north of the equator and most of the precipitation in the country is therefore caused by the polar front where dynamic low pressures are formed and warm air rises above colder air. The prevailing wind comes from the west, thus bringing humid air masses from the North Atlantic and the North Sea to the country. Most of the precipitation falls over the western part of the country (about 800-900 mm per year) whereas Copenhagen, which is located in the eastern parts of the country, only receives about 600 mm year [1]. However, the heavy rains which Copenhagen has experienced in later years were not caused by the processes normally causing precipitation in Denmark. They occurred during the summers when already hot and humid air were locally warmed, hence causing the air to rise and the water vapor in the rising air to condense and precipitate. The amount of precipitation is affected by the local heating, the absolute and relative humidity and the amount of aerosols in the atmosphere. Cities are characterized by both local heating and a relatively high amount of aerosols in the atmosphere around which the water vapor can condense.

The Danish area (except Bornholm) has for many million years been part of a sinking area, resulting in many different sediments beings deposited in the area. Thus, the underground consists of various sedimentary layers. Due to transportation of Tertiary sediments towards the western part of the country and the North Sea, the layer right below the Quarternary sediments in the eastern part of the country (including Copenhagen), consists of limestone, which is an excellent aquifer containing groundwater. Groundwater is also found in the Quarternary layer, but the main reservoirs are located in the limestone just below. During the last ice age, the ice covered the eastern part of the country. This resulted in deposition of particles of mixed size, including great amounts of clay, in the eastern part of the country. These depositions have great influence on the amount and quality of groundwater. The small pore spaces result in low infiltration velocity and the high field capacity together with the low infiltration velocity result in low percolation. Thus, in the eastern part of the country, formation of groundwater is relatively slow, but the slow infiltration velocity and the high amount of clay result in better purification of the groundwater. However, in Copenhagen a great use of groundwater has lowered the water table, and pollution from different sources on the ground has also contributed to lower quality of the groundwater.

We must assure that we produce groundwater, but concurrently the water must be uncontaminated. The groundwater is used for drinking, but for many other purposes as well (more than 99 % of the total amount of water consumption in Denmark comes from groundwater) [1]. Besides, most of the water running in streams comes from the groundwater, and groundwater also runs underground towards the sea/harbor.

It is important to secure the amount of groundwater for other reasons as well. Lowering the water table cause construction fundaments usually supported by the pressure of the groundwater to be more fragile, an issue which especially concerns the inner part of Copenhagen, where houses have been built on poles. Lowering the water table will also result in formation of Nickel and Arsenic, as well as salty groundwater moving inland. However, elevation of the water table can cause too much pressure on the fundaments, and infiltration of groundwater into basements and sewerage systems. Consequently, it is important to assure the current level of the water table. As a consequence of more evaporation and water running on the surface (due to more heavy rain) in the future, it is expected that the formation of groundwater will decrease. Hence, methods must be applied which increases the formation of groundwater and/or decreases the use of groundwater. However, some areas are more suitable for percolation of rainwater to the groundwater than others.

In Copenhagen, the sewerage system has been constructed in order to handle both waste water and rainwater. Waste water and rainwater are transported in the same pipes to the treatment plants where an advanced mechanical, chemical and biological treatment of the water makes it suitable for discharge to the harbor. The rainwater does not need the same comprehensive treatment, which means that unnecessary energy is used at the treatment plant. However, the coexistence of waste water and rainwater in the same pipes of the sewerage system also has other environmental impacts. During heavy rain, the system leads the water to the streams and the harbor untreated, which of course has a great impact on these natural environments. The capacity of the system has been improved and this scenario is therefore rare. During very heavy rain, the discharge of wastewater to streams and the harbor is not sufficient to discharge all the water from the sewerage system, resulting in waste water and rainwater flooding streets and houses. Separating rainwater from waste water would make the treatment of waste water at the treatment plant more energy efficient as well as preventing waste water from flooding streets, parks, houses etc. Numerous measures have been suggested and to some extend applied in order to secure this. Focus has been directed towards handling the rainwater locally, which can include storage, delay, transportation, and percolation. The local handling of rainwater is often designed as green areas. The question will remain though, whether or not it is efficient enough with green areas for drainage. The green areas will at some point become saturated by heavy rain, and as a result the water will run elsewhere. Furthermore, in some densely built areas of the city it is almost impossible to apply surface solutions. These areas will, if necessary, be augmented with piping, in which wastewater will be divided from rainwater. This piping shall not be as comprehensive as the standard piping, but instead be localized, and thereby maximized for its location, so that the unpolluted water will be lead to streams and the harbor in the most efficient way. This means that we will be able to avoid health violation risks such as coli bacteria spreading among the inhabitants in Copenhagen.

The solutions to higher frequencies of floods caused by heavy rain can in different ways affect the quantity and quality of groundwater and surface water, but also other environmental issues.

Purpose

The purpose of this study, is to contribute to the discussion about the most suitable solutions to higher frequencies of floods in Copenhagen with regard to the number and magnitude of floods, quantity and quality of ground water and surface water, and other environmental problems such as climate change and air pollution.

Method

The research on solutions to floods caused by heavy rain in Copenhagen is still new. However, a lot of research has already been done. The research implies predictions of future climate [2,3,4], movement of surface water [4], localization of areas which are exposed to floods [2,3,4], localization of areas where floods due to heavy rain will make huge damage (expressed in DKK) [2,3,4], general assessment of which solutions fits in different areas [2,3,4], and suggestions about specific solutions [5]. The methods applied includes IPPC's models predicting the future climate [2,3], predictions of consequences based on IPCC's scenarios for future climate [2,3,4], the URBAN MIKE method to predict surface water flows [2,3], and different tests of various solutions.

During our research we have found several studies already conducted. However, these studies have mostly focused on economically responsible solutions to reducing damages of floods. The environmental consequences of the solutions are mentioned only sporadically. Thus, our method has been to collect information from the various studies already conducted, and, using our knowledge about different environmental issues, come up with some suggestions for the best solutions. We do not suggest solutions for every street in Copenhagen. Instead we focus on the city as a whole, as well as specific areas of the city.

Results of the research

General principles

In Copenhagen, there is a general tendency that the western parts of the city are located above the eastern parts which are located next to the harbor. This results in surface water and groundwater flowing in an eastward direction. Nevertheless, local depressions and the built environment have made the flow directions of the surface water more complicated.

Additionally, exploitation of groundwater in Frederiksberg has made groundwater flowing towards the exploitation area (see fig. 1).

The reports which have been conducted suggest some general principles for solutions to floods. In higher altitude areas the water should be retained or slowed down (see fig. 2), whereas inclined and lower altitude areas should lead the water towards the harbor. However, solutions in higher altitude areas should also allow excess water to flow towards lower altitude areas and eventually the harbor (see fig. 2), and inclining and lower altitude areas should also be able to retain or slow down water outside the main water corridors.

Fig. 1: Blue curves represent equipotential lines. Groundwater flows perpendicular to equipotential lines towards areas where the groundwater potential is lower, which in this case will be towards the five groundwater drillings represented by the blue dots, and to a lesser degree the seven preventive groundwater drillings represented by red dots. The grey northwest-southeast zone represents the Carlsberg Fault, which contains limestone which is very permeable and therefore especially suitable for exploitation.

We have added the suggested solution to the map: The green line represents the location of the reopened Ladegårds Stream. The black line represents the location of the suggested tunnel. The orange line represents the location of the suggested stream running from Skt. Jørgens Sø to the harbour. The dotted orange line represents a solution where water from Ladegårds Stream continues directly in a water tunnel which goes towards the harbour in the south. The dotted orange line is an alternative solution, which should be applied if investigations show that Skt. Jørgens Sø can be polluted during floods and/or the ecosystem would suffer considerably from lowering the water level in the lake.

Source: [6]

Fig. 2: Skt. Kjelds Plads is the center of a relatively high altitude area in Copenhagen. The figure shows the appearance of Skt. Kjelds Plads after a huge amount of asphalt has been removed and substituted with green and blue elements. Skt. Kjelds Plads and its surrounding area is an example of an area which, during heavy rain, should retain and slow down water as well as leading excess water to lower altitude areas and eventually the harbor. If the small park surrounded by the roundabout is slightly lowered it would be able to retain water. Canals could also be constructed, in some cases by designing bikeways so they could function as canals during heavy rain.

Source: [7]

The transport of surface water towards the harbor should be concentrated in certain big corridors, and nearby roads should lead the surface water towards the corridors as well as absorbing it and slowing it down. These general principles are considered to be the best in order to prevent floods. A number of streets have been suggested as water corridors which should transport the surface water towards the harbor. One of them is Åboulevarden which is a main road leading traffic through the city. Another one goes along the western side of the lakes continuing towards the south to end up in the harbor. Development of these corridors in order to improve their ability to handle water is therefore necessary. Fig. 3 and 4 (see next subsection) show that the corridors suggested will drain big areas of the city which are highly vulnerable to floods.

Åboulevarden has been built by placing a former manmade stream (Ladegårds Å) in an underground pipe and constructing the road above, in some places elevated above the surrounding terrain. Several ideas have been suggested for appropriate solutions concerning Åboulevarden.

Suggested solution

Figure 3 and 4 below show our suggested solution.

Fig. 3: Blue arrows show direction of surface water. The green line represents the location of the reopened Ladegårds Stream. The black line represents the location of the suggested tunnel. The orange line represents the location of the suggested stream running from Skt. Jørgens Sø to the harbour. The dotted orange line represents a solution where water from Ladegårds Stream continues directly in a water tunnel which goes towards the harbour in the south. The dotted orange line is an alternative solution which should be applied if investigations show that Skt. Jørgens Sø can be polluted during floods and/or the ecosystem would suffer considerably from lowering the water level in the lake.

We have added the green, black, and orange lines to a map which have been constructed for another similar solution. However, the location and direction of the blue arrows are the same for the two solutions.

Source: [4]

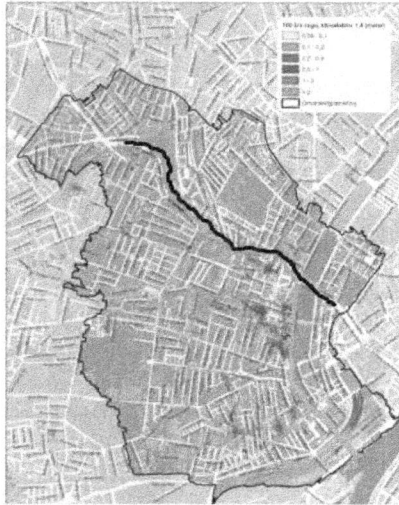

Figure 4: The colours indicate calculated floods (in meters) during heavy rain which statistically only occurs every 100 years, if no solutions are applied in order to prevent floods. The green line represents the location of the reopened Ladegårds Stream. The black line represents the location of the suggested tunnel. The orange line represents the location of the suggested stream running from Skt. Jørgens Sø to the harbour. The dotted orange line represents a solution where water from Ladegårds Stream continues directly in a water tunnel which goes towards the harbour in the south. The dotted orange line is an alternative solution which should be applied if investigations show that Skt. Jørgens Sø can be polluted during floods and/or the ecosystem would suffer considerably from lowering the water level in the lake.
Source: [5]

In general, we support the solution suggested by Miljøpunkt Nørrebro [8]. This solution implies exposure of the stream and construction of a tunnel (similar to the SMART tunnel in Kuala Lumpur) which, through two pipes, will lead traffic through the city. During heavy rain the stream will transport water towards the harbor via Skt. Jørgens Sø which is a lake, and if necessary, one of the pipes will be closed for vehicles and opened for water. Studies concerning the amount of water flowing and the hydraulic capacity of the stream combined with experiences in Kuala Lumpur, indicate that this solution is able to prevent over 10 cm of water on terrain from heavy rain, which statistically only occur every 100 years [8]. Other positive effects of the solution include increased recreational value, no pollution of the stream from vehicles (not different from today), no pollution of the ground from vehicles, and decreased pollution of the air in the area.

Construction of a tunnel in Åboulevarden will create space for a bigger green area around the stream. This area can be used for overflow during heavy rain, planting of vegetation which is tolerant to heavy rain and drought, and surface solutions which allow the water to percolate to the groundwater. Regarding percolation, infiltration beds (see fig. 5) and fascines like the BIO-BLOK model would be ideal. Infiltration beds should be placed in hollows in the terrain which could be nearby the stream. The main water table is located approximately 10 meters under the terrain

in Åboulevarden and groundwater will not enter the stream but move towards the groundwater drillings (see fig. 1). Exceptions to this could be where perched water tables exist.

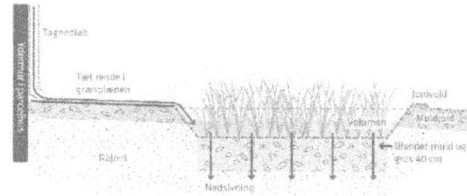

Fig. 5: The infiltration bed has vegetation and a top soil which fits the vegetation and is capable of retaining phosphorus, heavy metals, etc. Below is a layer of sand and gravel which allows fast infiltration.
Source: [10]

Behind the infiltration bed, construction of a small embankment could retain the rainwater for percolation, but the construction should allow excess rainwater to flow towards the stream. The infiltration beds should contain plants which can tolerate both periods without rain and periods with great amounts of rain. The infiltration beds often contain water because the water is flowing towards them during rain. However, during periods without rain, the layer of sand/gravel and the fascine below will cause dry conditions in the soil.

The infiltration beds are ideal for storage and percolation of rainwater, but at the same time they can provide aesthetic value to the area as well as improving living conditions for insects and other wildlife.

BIO-BLOK® fascines have a huge total surface area where microorganisms are growing. These microorganisms decompose organic pollutants. The fascine functions as a small biological treatment plant.

We suggest that the construction of the Ladegårds Stream should resemble a natural creek without concrete. The stream and its surroundings should contain plants, preferably of great variability, some of which should be hyperaccumulators (see fig. 6).

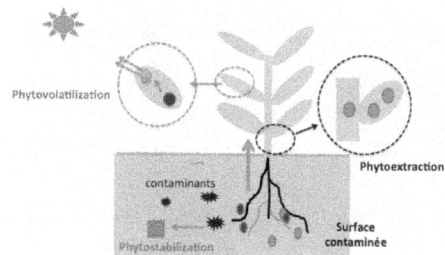

Fig. 6: Hyperaccumulators have a natural ability to bioaccumulate, decompose or render harmless pollutants in soil and groundwater. During growth, the roots of the plants absorb the metals from the soil. The metals are then stored in the plant. Examples of these plants are Vallisneria americana (Eelgrass), which remove Cd(H) and Pb(H), Azolla filiculoides (Water Fern), which are able to remove Ni(A), Pb(A) og Mn(A), and Helianthus annuus (Sunflower).
Source: [9]

Stones and a great variability of plants will create good living conditions for animals and reduce the amount of pollutants in the water. The green character of the creek will contribute to a more integrated system of green areas. Furthermore, vegetation will prevent very high temperatures due to transpiration.

Different solutions have also been suggested concerning the continued flow of the water from the Ladegårds Stream.

The solution with lowering the water level in Skt. Jørgens Sø and make it function as a storage bassin for water coming from the Ladegårds Stream could cause more or less polluted water to enter the lake during floods. However, the water flowing from the lake to the harbor secures that at least some of this potential pollution will be removed. It is then necessary to secure a proper treatment of the water before it flows into the harbor. Lowering the water level in Skt. Jørgens Sø would allow establishment of green areas around the lake. This could provide yet another piece for creating an integrated system of green areas in the city, thus improving conditions for bicycles and pedestrians as well as different species of plants and animals. The environment in and around the lakes will be further improved when the SMART tunnel continues under the lakes.

Furthermore, like everywhere else, construction of green areas will be able to decrease pollution of the atmosphere, including the level of CO_2.

Hence, we suggest a solution which includes a lowering of the water level in Skt. Jørgens Sø. However, the before mentioned risk of polluting the lake during floods and the consequences for the ecosystem by lowering the water level should be taken into account. Should future investigations show that the lake is highly vulnerable to pollution during floods and/or that the ecosystem would suffer considerably from lowering the water level, another solution should be applied instead. This solution implies that the water from the Ladegårds Stream and the tunnel below, instead of running through the lake, will run directly into and through a water tunnel which ends up in the harbor. However, this solution would make substitution of the water in the lake difficult.

We suggest that the water flowing in the Ladegårds Stream should continue through a tunnel which extends all the way to the harbor in the south. The corridor between the inner lakes and the harbor is very densely built. This makes it difficult to apply the same kind of solutions as in Åboulevarden. Placing the stream underground would make it less vulnerable to pollution.

The preceding sections have especially focused on quantity and quality of groundwater and surface water. However, other environmental aspects should also be included in the discussion of appropriate solutions. According to a research project carried out by Rambøll [8], the tunnel-solution suggested in this report will increase the traffic on Åboulevarden, while adjacent streets, which are now overloaded with traffic, will experience less traffic and thereby less congestion. The effect of the tunnel-solution on the total amount of vehicle traffic is difficult to assess. Considering that the total amount of vehicle transport will not be changed, the fact that the tunnel-solution will result in less congestion, will cause less pollution. The tunnel will also prevent pollution to spread to the atmosphere and the soil. Furthermore, the decreased emission of aerosols to the atmosphere will cause less condensation resulting in less precipitation. This could reduce the number and magnitude of heavy rain. The tunnel will not be able to prevent spreading of CO_2 to the atmosphere. However, construction of a green area along the creek above the tunnel will, like construction of green areas elsewhere, absorb CO_2 from the atmosphere. It can also provide better conditions for bicycles and pedestrians, which can result in less traffic by cars.

Furthermore, it will result in a smaller amount of pollutants in the atmosphere, the soil, and the groundwater, as well as having a positive effect on the biodiversity, and a stabilizing effect on temperatures.

Conclusion

Based on research already carried out, we have suggested solutions to floods which we think combine the two aims of reducing floods and securing a sustainable environment. For Åboulevarden, we suggest a solution where the stream is brought back to the surface and the highway is placed underground. This solution would have various positive effects, including: Huge transporting and storage capability of water during floods, formation of groundwater, natural water treatment, more green and blue areas, and less congestion. For the area between Ladegårds Stream and the harbor to the south, we suggest a lowering of the water table in Skt. Jørgens Sø, and a connection between the stream and the lake as well as a connection between the lake and a tunnel which transports the water towards the harbour. The lowering of the water level in the lake would result in great storage capacity during floods as well as a green area which benefits the environment in various ways. However, improved knowledge about the extent of possible pollutions of the lake during floods, and possible effects on the ecosystem, will provide knowledge which can improve the basis from which decisions should be taken. Based on these possible environmental risks, we have suggested an alternative solution where the water running in the Ladegårds Stream and the tunnel below continues to the harbour in a tunnel without affecting the lake. Another relevant study to be carried out could be further investigation of the effects of different solutions on traffic, which also include the effects of the new metro city ring and other means of public transport, which could be introduced.

Acknowledgements

We are grateful to Martin Drews, Per Skougaard Kaspersen, and Simon Bolwig from the Technical University of Denmark for taking time to discuss our project with us. We would also like to thank our teachers, Anne-Mette Chistiansen and Lars Rostgaard Toft for their guidance throughout the project.

References

[1] Sanden, E et.al. (2005). *Alverdens Geografi*, GEOGRAFFORLAGET, pp. 65-81

[2] Københavns Kommune (2011). *Københavns Klimatilpasningsplan*. Formula. pp. 1-99. http://www.kk.dk/da/om-kommunen/indsatsomraader-og- politikker/natur-miljoe-og-affald/klima/klimatilpasning

[3] Frederiksberg Kommune (2012). *Klimatilpasningsplan 2012*. pp. 1-66. http://www.frederiksberg.dk/Politik-og-demokrati/Politikker-og-strategier/Miljoe-klima-og-affald.aspx#B076A2A0F511441EB347FC42CDF2AAF2

[4] Københavns Kommune (2012). *Skybrudsplan 2012*. pp. 1-32. http://www.kk.dk/da/Om-kommunen/Indsatsomraader-og-politikker/Publikationer.aspx?mode=detalje&id=1018

[5] Københavns og Frederiksberg Kommuner (2013). *Konkretisering af Skybrudsplan – Ladegårdså, Frederiksberg Øst og Vesterbro*. pp. 1-14. https://subsite.kk.dk/~/media/0ABB6FADE4124980B240746BF2D5F551.ashx

[6] Frederiksberg Kommune (2009). *Grundvandsplan for perioden* 2009-2010, pp. 1-109

[7] Københavns Kommune (2014). *Skt.Kjelds Plads* http://www.klimakvarter.dk/byrum/skp/

[8] Jensen, A.J. (2014) : *Åbn Åen* http://www.ladegaardsaaen.dk/

[9] https://lh3.googleusercontent.com/-JjsVNAT4XcA/UTXvSUQ7CQI/AAAAAAAAFa8/WBB-vuUip0Y/s667/Slide1.gif

[10] http://www.klimatilpasning.dk/media/6454/regnbed.jpg

Removal of Heavy Metal Ions Using Durian Waste

Wei Heng Tan, Bryan Lim, Ruobing Han

Hwa Chong Institution, Singapore, pehyk@hci.edu.sg

Abstract

Industrial wastewater that is being discharged into water has been causing a multitude of problems environmentally. One of the reasons for this is due to the presence of heavy metal ions which are toxic and known to bioaccumulate in living organisms, causing detrimental effects. This study aims to investigate the ability of durian wastes – husk, rind, outer seed and inner seed in adsorbing copper(II), iron(III) and lead(II) ions. The metal ion adsorbed onto the durian waste was desorbed using nitric acid and the reusability of the desorbed durian waste investigated. Fourier Transform Infrared Spectroscopy (FTIR) revealed the presence of carboxyl and hydroxyl groups in the durian waste which are likely to be responsible for the adsorption of metal ions. The husk is most effective in the adsorption of lead(II) ions while the outer part of seed is most effective in the adsorption of iron(III) ions. Husk, rind and outer part of seed are comparable in terms of their ability to adsorb copper(II) ions. The inner part of seed is the least effective in the adsorption of all 3 types of metal ions. Husk, rind and outer part of seed can be reused and regenerated for at least 3 cycles.

Keywords

Adsorption, durian waste, desorption, regenerated

1 Introduction

Heavy metal pollution is one of the most important environmental problems today. Various industries produce and discharge wastes containing different heavy metals into the environment, such as mining and smelting of metalliferous, surface finishing industry, energy and fuel production, fertilizer and pesticide industry and application, metallurgy, iron and steel, electroplating, electrolysis, electro-osmosis and leatherworking [1]. Heavy metals pose a threat to the environment as they cannot be degraded and would remain in the soil and water for a long period of time. They produce their toxicity by forming complexes with proteins, in which carboxylic (-COOH), amine (-NH$_2$), and thiol (-SH) groups are involved. These modified biological molecules lose their ability to function properly and result in the malfunction or death of the cells. When metals bind to these groups, they inactivate important enzyme systems or affect protein structure, which is linked to the catalytic properties of enzymes [2].

Metal ions can be removed from industrial effluents through several methods. Many existing methods for treatment include chemical and surface chemistry processes such as precipitation, adsorption, membrane processes, ionic exchange and reverse osmosis [3]. However, chemical precipitation and electrochemical treatment are ineffective, especially when metal concentration in aqueous solution is among 1 to 100 mg/L, and also produce large quantities of sludge. Ion exchange, membrane technologies and activated carbon adsorption process are expensive when treating large amount of wastewater containing heavy metal in low concentration, and this means they cannot be used at large scale [4]. Hence there is a need to explore alternative ways to remove the heavy metal ions.

One such alternative is biosorption. Biosorption, which utilises inactive biological materials for the removal of heavy metals, is gaining popularity due to its low cost and environmentally friendly nature. Biomaterials, such as seaweed [5], crab shell [6] and soybean hulls [7] have been investigated for their biosorptive properties.

Durian (*Durio zibethinus*) is an exotic fruit which is popular in Singapore, Malaysia and South East Asian Countries. However the edible portion of the fruit, the flesh, is only 15-30% of the entire mass of the fruit, whereas the peel and seeds which constitute about 70-85% of the fruit are discarded [8]. Hence there is a need to explore ways to fully utilise the durian waste.

This study investigates the ability of durian wastes (husk, rind and seed) in removing copper(II), lead(II) and iron(III) ions. Durian wastes are chosen as they are under-utilised in Singapore and because they are easily available. In addition, agricultural by-products contain cellulose and lignin which possesses polar functional groups such as the hydroxyl, phenolic and carboxyl group [9]. These groups are known to bind to heavy metal ions by donation of an electron pair to form complexes with metal ions in solutions. The reusability of the durian waste after desorption was also studied.

2 Materials and Methods

2.1 Materials

Durian waste was collected from local fruit stores. Iron(III) chloride, lead(II) nitrate, and copper(II) sulfate were purchased from GCE Laboratory Chemicals. Nitric acid was purchased from Honeywell.

2.2 Preparation of durian wastes

Durian waste including peels and seeds were soaked in deionised water for one hour to remove any adhering dirt or durian. The peel was separated into husk and rind using a sharp knife. Both the husks and rind were cut into smaller sizes of 1 cm by 1 cm. Seeds were separated into outer part and inner part. The husk, peel and seeds were then dried in a hot air oven at 70 °C until constant mass. Then, the waste was ground, blended and sieved to ensure consistency in particle size.

2.3 Adsorption tests

Three solutions each containing 50 ppm of copper(II), lead(II) and iron(III) ion were prepared from copper(II) sulfate, lead(II) nitrate and iron(III) nitrate in deionized water respectively. 0.5 g of durian waste was added to 50ml of 50 ppm of heavy metal ion solution. The mixture was then stirred for one hour. A control set up with only the metal ion solution but without the durian waste in it was subjected to the same experimental condition. At the end of one hour, the mixture was centrifuged. The supernatant was analysed for the metal ion remaining using a colorimeter (Hach, DR 890) for iron(III) and copper(II) ion and an AA Spectrophotometer (AA 6300 Shimadzu) for lead(II) ion. Experiments were conducted in 5 replicates.

2.4 Regeneration and reusability tests

0.5 g of durian waste was added to 50 ml of 50 ppm of heavy metal ion solution. The mixture was stirred for one hour. The mixture was centrifuged and the supernatant was analysed to determine the amount of metal ions being adsorbed by the waste. The waste was then transferred to another beaker containing 50 ml of 0.05 M HNO_3. The mixture was stirred for one hour. It was centrifuged to separate the desorbed durian waste.

The desorbed durian waste was washed with deionised water and dried using an oven at 70°C until constant mass. The ability of desorbed durian waste in adsorbing copper(II) was carried out. Desorption and regeneration processes were repeated for a number of times to determine how many times the adsorbent can be reused. For each cycle, the mass of durian waste to volume of metal ion solution ratio was kept constant. After each cycle, the amount of metal ion which could still be adsorbed by the adsorbent was determined. Experiments were conducted in triplicates.

3 Results and Discussion

3.1 Adsorption tests

3.1.1 Lead(II) ions

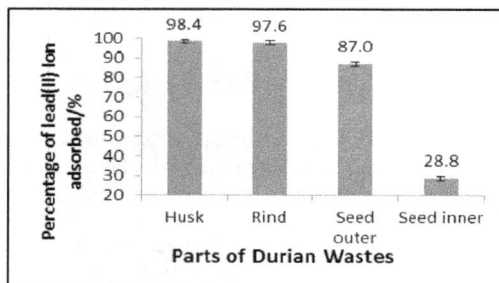

Figure 1: Adsorption of lead(II) ion by durian waste

Figure 1 compares the ability of different durian waste in adsorbing lead(II) ion. Husk is most effective in adsorbing lead(II) ion, followed by rind, outer part of seed and inner part of seed. Interestingly, inner part of seed adsorbed less than 50% of lead(II) ion, unlike the other 3 types of durian wastes which were able to remove more than 80% of lead(II) ion.

3.1.2 Iron(III) ions

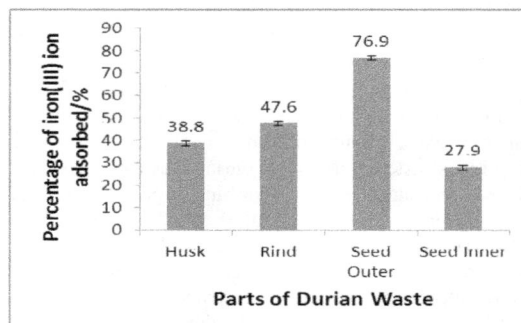

Figure 2: Adsorption of iron(III) ion by durian waste

Figure 2 shows that outer part of seed is most effective in adsorbing iron(III) ion, followed by rind, husk and inner part of seed. The percentage of iron adsorbed by durian waste is less than that of lead(II) ion. Again inner part of seed is observed to be least effective in adsorbing metal ions.

3.1.3 Copper(II) ions

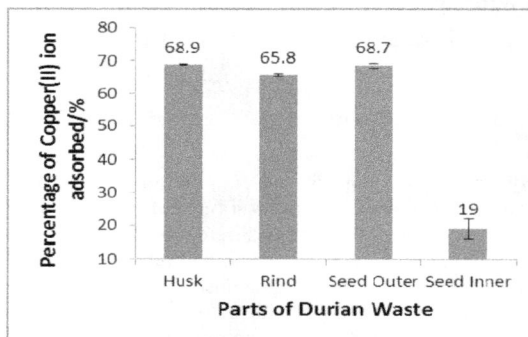

Figure 3: Adsorption of copper(II) ion by durian waste

Figure 3 shows that the husk, rind and outer part of seed is comparable in terms of the percentage of copper(II) ion adsorbed while the Inner part of seed is least effective in adsorbing copper(II) ions. The percentage of copper(II) ion adsorbed by durian waste was less than the percentage of lead(II) ion adsorbed.

Inner part of seed is least effective in adsorbing copper(II) ion. This observation was consistent with the adsorption of lead(II) and iron(III) ion.

3.2 Characterization of durian waste

Figure 4 shows the FTIR spectrum of durian husk. The FTIR spectrum displayed a number of peaks, indicating the complex nature of the adsorbent. The spectrum displayed the following bands: 3407 (O-H stretching vibrations), 2927(C-H stretching vibrations), 1731 (C=O stretching vibrations of carboxylic acid), 1635 (C=C stretching vibrations), 1375 (carboxylate anion stretching), and 1057 (C-O-H stretching) cm^{-1}.

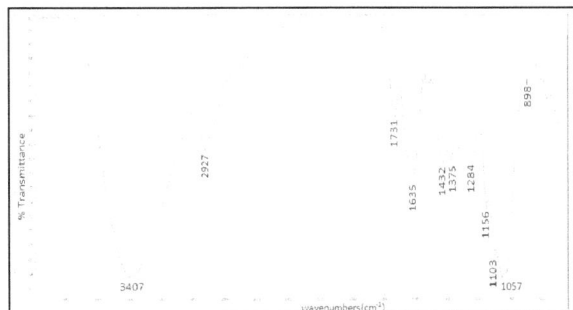

Figure 4: FTIR Spectrum of durian husk

Previous study has reported that heavy metal ions interact with various functional groups such as carboxyl, amino, thiol, hydroxyl and phosphate groups via the formation of co-ordinate bonds [10] Thus it is likely that the hydroxyl (-OH) group and the carboxyl group (-COOH) present in durian waste are responsible for the binding of metal ions.

Figure 5: Comparison of the FTIR spectra of different parts of durian waste

The spectra of husk, rind, outer part of seed and inner part of seed were compared. The result is shown in figure 5. All spectra look alike except that the band around 1730 cm^{-1} (C=O stretching) is not evident in the spectrum of inner part of seed. Carboxyl groups (-COOH) are known to bind to heavy metal ions [10]. Hence the absence of carboxyl groups in inner part of seed explains why it is the least effective in the adsorption of copper(II), iron(III) and lead(II) ions.

The spectra before adsorption were compared with that after adsorption. It was observed that the bands corresponding to C=O and OH groups have shifted (table 1). This further supports the hypothesis that these groups are responsible for the adsorption of metal ions.

Table 1: Comparison of FTIR peaks before and after adsorption

Type of durian waste	Type of metal ion	Peak	Wavenumber before adsorption (cm^{-1})	Wavenumber after adsorption (cm^{-1})
Rind	Pb^{2+}	O-H stretch	3408	3422
		C=O stretch	1731	1739
Seed outer	Cu^{2+}	O-H stretch	3376	3417
		C=O stretch	1727	1729
Seed outer	Fe^{3+}	O-H stretch	3376	3381
		C=O stretch	1727	1730

3.3 Regeneration tests

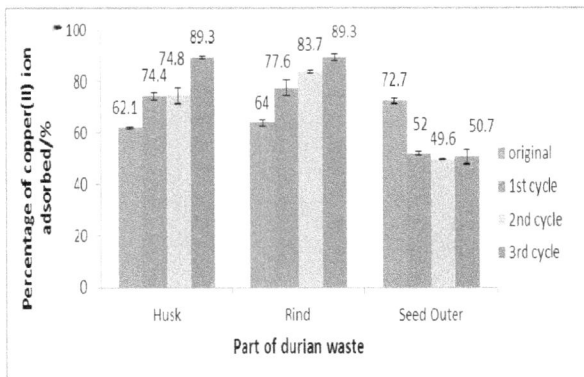

Figure 6: Amount of copper(II) ion adsorbed after each cycle of desorption

Figure 6 shows the amount of copper(II) ion adsorbed after each cycle of desorption of different part of waste. The study on inner part of seed was discontinued as it was least effective in adsorbing all three types of metal ions. The percentage of copper(II) ion adsorbed increases as the number of desorption cycle increases for both husk and rind. Desorption of durian waste was carried out with dilute nitric acid. Previous study has shown that treatment of cellulosic waste with dilute acid can achieve reaction rates and improve cellulose hydrolysis, thus exposing the active sites for binding of metal ion to occur [11]. This explains why ability of metal ion adsorption increases with each cycle of desorption with dilute nitric acid.

However outer part of seed did not show the same trend. There is a drastic drop in the percentage of copper(II) ion adsorbed after first cycle of desorption but the decrease is less drastic for subsequent cycles of desorption.

Durian waste can be reused and regenerated for at least 3 cycles. However it was observed that the mass of the durian waste decreases with each cycle of desorption. At the end of 3rd cycle, mass of remaining waste is too small to carry on further regeneration.

4   Conclusion

Durian husk, rind and seed are able to adsorb copper (II), lead (II) and iron (III) ions. Among the 3 types of durian waste, the husk is most effective in the adsorption of lead (II) ions while the outer part of seed is most effective in the adsorption of iron (III) ions. Husk, rind and outer part of seed are comparable in terms of the ability to adsorb copper(II) ions. The inner part of seed is the least effective in the adsorption of all 3 types of metal ions, due to the absence of carboxyl groups.

Durian waste can be reused and regenerated for at least 3 cycles, with the husk and rind showing an increase in the percentage of copper(II) ion adsorbed after each cycle of desorption. The outer part of seed however experienced a drastic drop in the percentage of copper(II) ion adsorbed after first cycle of desorption but subsequently the drop in percentage of copper(II) ion adsorbed was less drastic.

According to a study conducted by Duazo [12], durian seeds were found to exhibit antibacterial property against *Escherichia coli* and *Staphylococcus aureus*. Coupled with the fact that durian waste are able to remove metal ions, it would be interesting to construct a filter using durian waste and investigate its effectiveness in removing both metal ions and bacteria. Such a filter could be used to purify water in countries such as Thailand and Cambodia where durians are abundant.

Acknowledgements

We would like to thank Mrs Sow Yoke Keow and Mdm Xia Ying for their guidance.

References

[1]   Wang, J. L., Chen, C.(2006). Biosorption of heavy metals by Saccharomyces cerevisiae: A review. *Biotechnol Adv,* 24, 427-51.

[2]   Adepoju-Bello, A.A, and Alabi, O.M. (2005). Heavy metals: A review. *The Nig. J. Pharm.* 37, 41-45.

[3]   Saikaew, W., Kaewsarn, P. (2010). Durian Peel as biosorbent for removal of cadmium ions from aqueous solutions. *J. Environ. Res.*, 32(1), 17-30.

[4]   Volesky, B. (2001). Detoxification of metal-bearing effluents: biosorption for the next century. *Hydrometallurgy*, 59, 203-16.

[5]   Davis, T.A., Volesky, B., Mucci, A. (2003). A review of the biochemistry of heavy metal biosorption by brown algae. *Water Res.*, 37, 4311-4330.

[6]   An, H.K., Park, B. Y., Kim, D. S. (2001). Crab shell for the removal of heavy metals from aqueous solution. *Water Res.,* 35, 3551-3556.

[7]   Marshall, W. E., Wartelle, L.H., Boler, D. E., Johns, M. M., Toles, C. A. (1999). Enhanced metal adsorption by soybean hulls modified with citric acid. *Biores. Technol.* 69, 263-268.

[8]   Mohammed, S.A., Najib, N. W.A.Z., & Muniandi, V. (2012). Durian rind as a low cost adsorbent, International *Journal of Civil & Environmental Engineering IJCEE-IJENS*, 12, 04.

[9]   Pagnanelli, F., Mainelli, S., Veglio, F. and Toro, L. (2003). Heavy metal removal by olive pomace: biosorbent characterization and equilibriummodeling. *Chem. Eng Sci.*, 58, 4709-4717.

[10]   Sag, Y., Kutsal, T. (2001). Recent Trends in the Biosorption of Heavy Metals: A Review. *Biotechnol Bioprocess Eng*, 6, 376-385

[11]   Esteghlalian, A., Hashimoto, A.G., Fenske, J.J., Penner, M.H. (1997). Modeling and optimization of the dilute-sulfuric-acid pretreatment of corn stover, poplar and switchgrass. *Bioresour. Technol.*, 59, 129–136.

[12]   Duazo, N.O., Bautista, J. R., Teves, F. G. (2012). Crude Methanolic Extract Activity from Rinds and Seeds of Native Durian (Durio zibethinus) against Escherichia coli and Staphylococcus aureus. *African Journal of Microbiology Research*, 6(35), 6483-6486.

Optimization of PVDF membrane

Daniel Yang, Ng Xiang Han, Kenrick Kwa Zheng Feng, Neville Cheong

Hwa Chong Institution, Singapore, kongcw@hci.edu.sg

Abstract

Membrane technology is nowadays increasingly used as an alternative to wastewater treatment. Our membrane is made of PVDF, PVP and PEG using DMAC as solvent. The project aims to find out the effect of different percentages of PVP and PEG on the PVDF membrane in term of pore size, tensile strength, water permeation and removal of particulate matters. As the percentages of PVDF, PVP and PEG increase, the average pore size and hence the water permeation of the membrane decrease. PEG had less effect on the average pore size and water permeation as compared to PVDF. An increase in percentage PEG from 12 % to 20 % led to 25 % reduction in water permeation as compared to 86 % reduction for a similar increase in percentage PVDF. The membrane tensile strength increases as the percentages of PVDF, PVP and PEG increase. Our results showed that an increase in percentages of PVDF and PVP led to an increase in bubble point; while an increase in percentage PEG caused a decrease in bubble point (used to measure the largest pore size). 10 % PVP was found to be the best composition in term of lowering turbidity with a reasonable water permeation rate. A portable filtration set was developed using hollow fiber membrane made of 10 % PVP.

Keywords

PVDF, PEG, PVP, bubble point, pore size, turbidity

1   Introduction

Membrane technology is increasingly used as an alternative to industrial wastewater treatment. Its application to water treatment has increased dramatically in the past decade with the improvement in membrane quality and the decrease in membrane costs. In particular, low pressure membrane technology has experienced accelerated growth. Drinking water treatment and wastewater reuse accounted for 82 % of the total capacity of low pressure membrane[8]. PVDF is a type of low pressure membrane. The structure of the membrane can be modified by using PVP and PEG. Structural formulae of the polymers are shown below.

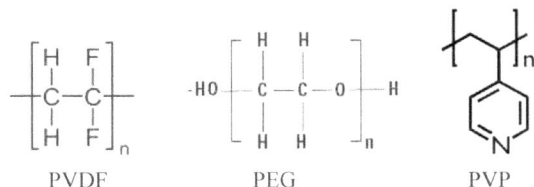

Poly(vinylidenefluoride) (PVDF) (molecular mass 360,000) is a highly non-reactive polymer, which is strong, and resistance to solvents, acids, bases and heat. It is used as the main component in a PVDF membrane. Like PVDF, poly(vinylpyrrolidone) (PVP) (molecular mass 300,000) is also a highly non-reactive polymer, which helps to improve the hydrophilicity of the membrane, i.e. the water flux (water permeation). Poly(ethyleneglycol) (PEG) on the other hand is an aqueous polymer solution (molecular mass 400) which is highly polar. It aids the pore formation of the membrane[4]. Dimethylacetamide (DMAC) is used as a solvent in the fabrication of the membrane. Both PVP and PEG are added as additives to modify the structure and hence the characteristic of the membrane.

In this research project, we made a PVDF membrane and studied the effect of different composition of PVP and PEG on the performance of the membrane. We aimed to develop membrane with suitable composition that produces clean water with acceptable turbidity level. We also developed a portable filtration set using the membrane. The filtration set can be used to filter river water free of heavy metal ions and other soluble organic chemicals. It would be useful to people in less developed countries that have a limited access to clean and potable water. PVDF membrane is used to remove large particulate matters and it is not intended to remove small particles such as heavy metal ions and other soluble organic substances.

2.   Materials and Methods

2.1   Materials

Chinese calligraphy ink was purchased from bookshop. Poly(vinvlidenefluoride) (PVDF), poly(vinylpyrrolidone) (PVP), poly(ethyleneglycol) (PEG) and dimethylacetamide were purchased from GCE Laboratory Chemicals.

2.2   Fabrication of Membrane

PVDF, PVP and PEG were mixed with the solvent DMAC in a 80 °C water bath with constant stirring until all components dissolve. The compositions of the chemical used were varied from 12-20 % PVDF, 2-10 % PVP and 12-20 % PEG. The hot solution was poured on a white tile and was spread out evenly with a piece of glass slide. The white tile was then immersed in 25 °C deionised water. The solution on the white tile solidified immediately in water. The membrane was then rinse with deionised water for 10 minutes to remove the solvent.

2.3 Characterization of membrane

2.3.1 Average pore size

The pore size of the membrane was studied using scanning electron microscope (SEM) (Figure 1). The pore size was measured and compared with the scale. The average pore size was calculated by measuring the pore size in a selected region in the SEM image as shown in Figure 1 and Figure 2.

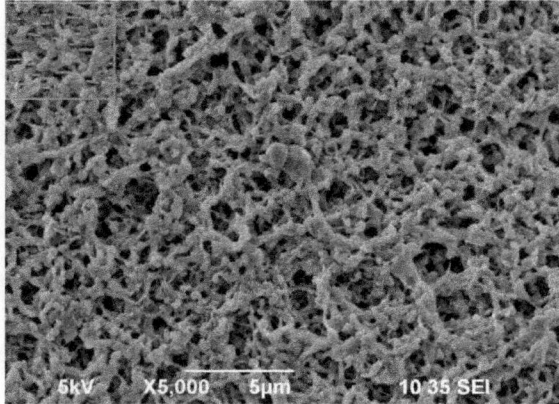

Figure 1: SEM image of the membrane.

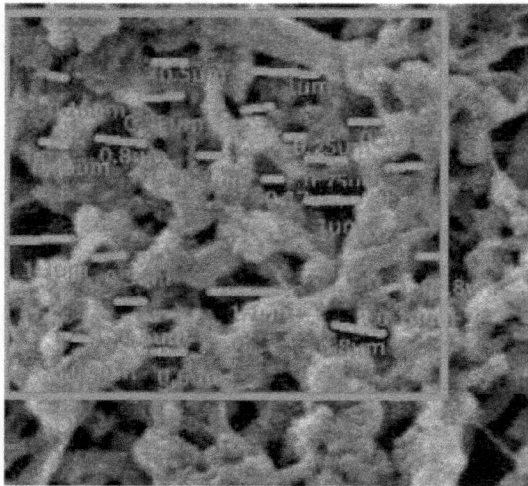

Figure 2: Selected region of the SEM image of the membrane (from Figure 1).

2.3.2 Water permeation

20 ml of water was filtered through the membrane. The time taken for water to pass through the membrane was measured. The water permeation of the membrane was then calculated using the formula below.

$$\frac{\text{Volume of water permeated (L)}}{\text{Surface area of the membrane (m}^2) \times \text{filtration time (h)}}$$

2.3.3 Bubble point test

The membrane was fixed on the holder of a pressure gauge and was then soaked in water. The pressure was slowly increased until the first bubble appeared through the membrane as shown in Figure 3. The minimum pressure to squeeze the first bubble through the membrane is known as bubble point. It measures the largest pore size of the membrane and hence determines the largest particle that can pass through the membrane.

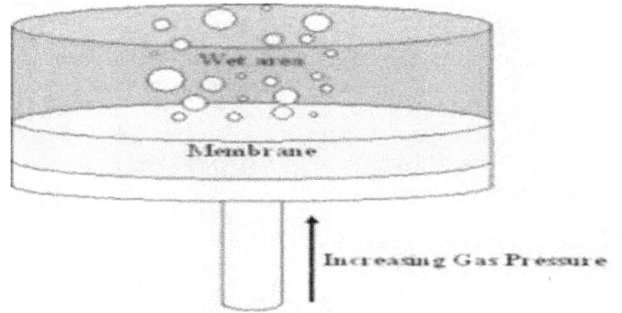

Figure 3: Bubble point test

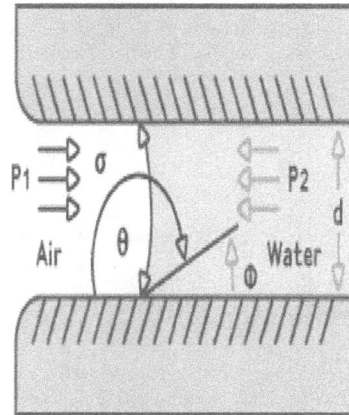

Figure 4: A cross sectional diagram of a pore in the membrane

Figure 4 shows a cross sectional diagram of a pore in a PVDF membrane. The pressure exerted at the dry region of the membrane will force the air bubble to pass through the membrane. The larger the pore size the smaller is the pressure required to squeeze the air through the membrane.

The largest pore size of the membrane can be calculated using the following formula [7]:

$$P = \frac{4k \cos \theta}{d} \sigma$$

Where:

P = bubble point pressure
d = pore diameter
k = shape correction factor = 1 (for cylindrical-shaped pore)
θ = liquid-solid contact angle = 0 (for water)
σ = surface tension = 72 mN m^{-1} (for water)

2.3.4 Tensile strength

The membrane was stretched till it broke. The minimum force required to tear the membrane was recorded as tensile strength of the membrane.

2.4 Water turbidity test

Chinese calligraphy ink was used to prepare water with turbidity of 9 NTU. The water was then filtered through the membrane. The turbidity of the filtrate was measured using turbidity meter.

3 Results and Discussion

3.1 Average pore size

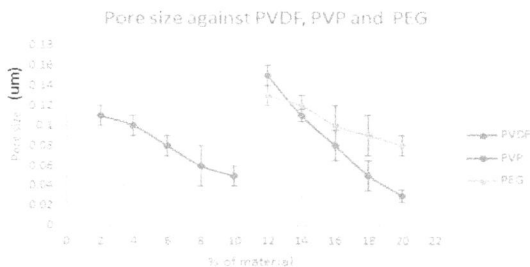

Figure 5: Average pore size of the membrane

Figure 5 shows the average pore size of the membrane with different composition of polymers. Generally, the average pore size decreases with increasing composition of PVDF, PVP and PEG. These polymers take up more space in the membrane and lead to a decrease in the average pore size. PVDF is the main component in the PVDF membrane. PEG however, decreases the average pore size to lesser extent as compared to PVDF. It is interesting to note that lower composition of PVP is needed to lower the average pore size as compared to PVDF and PEG.

3.2 Water permeation

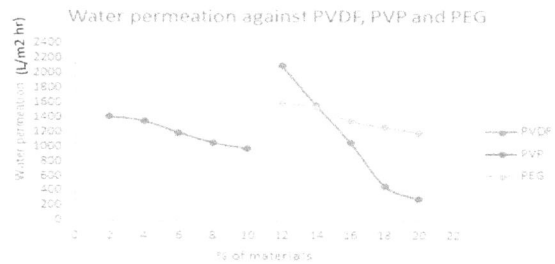

Figure 6: Water permeation of the membrane

Figure 6 shows the water permeation of the membrane with different composition of the polymers. Like average pore size, the water permeation of the membrane decreases with increasing composition of PVDF, PVP and PEG. This is consistent with the decreased average pore size as the percentages of the polymers increases (as shown in Figure 5). PVDF has the greatest effect on water permeation of the membrane, with a recorded 86 % reduction in water permeation. PVP and PEG were shown to cause mild reduction in water permeation. This may due to the fact that both PVP and PEG are more polar than PVDF. Hence, both PVP and PEG allow water which is highly polar to pass through the membrane[4].

3.3 Bubble point and largest pore size

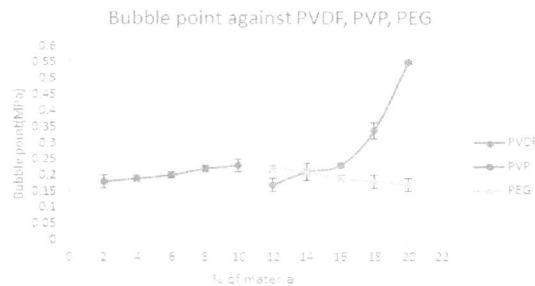

Figure 7: Bubble point test

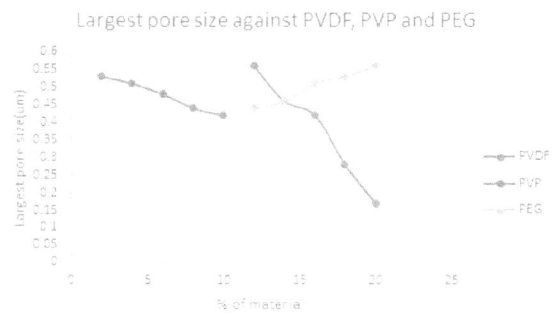

Figure 8: Largest pore size of the membrane calculated from bubble point test

Figure **7** shows the bubble point of the membrane with different composition of the polymers. Bubble point is the minimum pressure needed to squeeze the first bubble through the membrane. It measures the largest pore size in the membrane. The bubble points were converted to pore size using the formula shown in Methodology (Figure **8**). The larger the bubble point, the smaller the largest pore size. Hence, bubble point gives an indication of the largest particle size can pass through the membrane.

Both PVDF and PVP increase the bubble point of the membrane. This is consistent with the decreased in average pore size of the membrane as the composition of these polymers increases as shown in Figure **5**. However, PVDF increases the bubble point of the membrane exponentially as compared to PVP which only causes a slight increase in bubble point. Hence, it is easier to limit the size of the particle that can pass through the membrane by regulating the percentage of PVDF. However, higher composition of PVDF will cause a drastic decrease in water permeation, i.e. longer filtration time. Furthermore, high PVDF composition will produce brittle membranes.

It is interesting to note that the bubble point of the membrane decreases (i.e. larger pore size) with increasing percentage of PEG despite the fact that higher composition of PEG leads to smaller average pore size (as shown in Figure **5**). This could be due to the pore-forming property of PEG[4]. In other words, higher percentage of PEG increases the largest pore size but at the same time decreases the average pore size.

3.4   Tensile strength

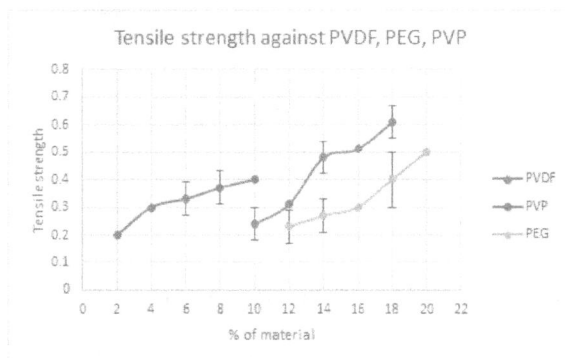

Figure **9**: Tensile strength of the membrane

Figure **9** shows the tensile strength of the membrane with different composition of the polymers. The tensile strength of the membrane increases with increasing PVDF, PVP and PEG. This is due to the cross-links formed between the polymers. PVDF strengthen the membrane more than PEG. But, higher PVDF composition will lead to lower water permeation rate. Furthermore, high PVDF composition will cause brittleness in the membrane. Hence, PEG and PVP can be used as modifiers to improve the membrane without compromising the tensile strength of the membrane.

3.5   Water turbidity test

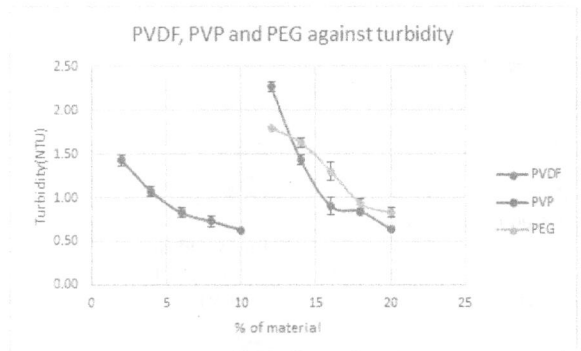

Figure **10**: Water turbidity test

Figure **10** shows the turbidity of water passing through the membrane with different composition. Chinese calligraphy ink was used to prepare the water sample with 9 NTU. All composition of polymers used in our research decrease turbidity significantly (about 95 % redcution) from an original value of 9 NTU. The Public Utilities Board (PUB) of Singapore requires all water treatment facilities to produce water containing less than 1 NTU[6]. All three polymers with the right composition are able to reduce the turbidity to less than 1 NTU.

10% PVP has similar turbidity with 20% PVDF and 20 % PEG. However, 10% PVP has higher water permeation (i.e. faster filtration) as compared to 20 % PVDF. It also has smaller average pore size and smaller largest pore size (i.e. limit the size of particle that pass through the membrane) as compared to 20 % PEG. Hence, 10 % PVP was chosen to make hollow fibers membrane with reasonable tensile strength and water permeation. These hollow fibers were then used to make a portable filtration set as shown in Figure **11**. The filtration set was shown to lower the turbidity of ink water from 9 to less than 1 NTU, which is within the standard set by Public Utilities Board (PUB) of Singapore[6].

Figure **11**: Portable filtration set with hollow fibers

4 Conclusion

As the percentages of PVDF, PVP and PEG increase, the average pore size and hence the water permeation of the membrane decrease. PEG has less effect on the average pore size and water permeation as compared to PVDF. An increase in percentage PEG from 12 % to 20 % led to 25 % reduction in water permeation as compared to 86 % reduction for a similar increase in percentage PVDF. However, an increase in percentage PVP form 2 % to 10 % led to 25 % reduction of water permeation, which is similar to PEG, but with smaller composition. The membrane tensile strength increases as the percentages of PVDF, PVP and PEG increase. Our results showed that an increase in percentages of PVDF and PVP led to an increase in bubble point; while an increase in percentage PEG caused a decrease in bubble point. Bubble point measures the largest pore size and hence determines the largest particle size that can pass through the membrane. High PVDF composition will ensure high membrane tensile strength with smaller average pore size and smaller largest pore size, which sacrifice water permeation rate. Hence, PEG and PVP can be used to increase water permeation rate of the membrane. 10 % PVP has similar water permeation rate as compared to 20 % PEG, but with smaller average pore size and smaller largest pore size. This is critical to as it limit the particle size that pass through the membrane. Hence, 10 % PVP was used to make hollow fiber membrane for the portable filtration set.

Acknowledgements

We would like to thank our mentor Mr. Kong Chiak Wu and Mdm Xia Ying for their guidance and invaluable advice.

References

[1] *Laizhou SONG, Zunju ZHANG, Shizhe SONG and Zhiming GAO. (2007). Preparation and characterization of the modified polyvinylidene fluoride (PVDF) hollow fibre microfiltration membrane. J. Mater. Sci. Technol. 23 (1), 55.*

[2] N. Awanis Hashim, Yutie Liu, K. Li. (2011). Stability of PVDF hollow fibre membranes in sodium hydroxide aqueous solution. *Chemical Engineering Science.* 66, 1565–1575

[3] Dan-ying Zuo, You-yi Xu, Wei-lin Xu, Han-tao Zou. (2008). The influence of PEG Molecular weight on morphologies and properties of PVDF Asymmetric membranes. *Chinese Journal of Polymer Science.* 26 (4), 405−414.

[4] Jian Chen, Jiding Li, Xia Zhan, Xiaolong Han, Cuixian Chen. (2010). Effect of PEG additives on properties and morphologies of polyethrimide membranes prepared by phase inversion. *Front. Chem. Eng. China.* 4 (3), 300-306.

[5] Xi Dan-Li and Zhou Yuan. (2008). Porous PVDF/TPU blends asymmetric hollow fiber membranes prepared with the use of hydrophilic additive PVP (K30) *Desalination* 223, 438-447.

[6] Public Utilities Board (PUB) of Singapore: www.pub.gov.sg.

[7] EMD Millipore: www.millipore.com

[8] Haiou Huang, Kellogg Schwab and Joseph G. J. Jacangelo. (2009). Pretreatment for low pressure membranes in water treatment: a review. *Environ. Sci. Technol.* 43 (9), 3011-3019.

Biosand Filters: Evaluation of Biochar augmented Kanchan filters for use in Cambodian Rural communities

Jonathan Seow, Joel Lee, Ong Jun Yi, Foo Ming Wei, Eunice Tan

NUS High School of Mathematics and Science, Singapore, nhsczm@nus.edu.sg

Abstract

In rural areas, safe drinking water is difficult to find, leading to many health problems especially in the young and the elderly. Biosand filters such as the *Kanshan Arsenic Filter* (KAF) have been shown promise in the removal of arsenic to acceptable standards, but only if arsenic concentrations are low enough. In addition, the KAF is unable to filter other heavy metals present in pesticides. Biocharcoal has been shown to be able to remove pesticides from drinking water. We hope to construct a filter with a combination of both systems in order to obtain drinking water with safe levels of pollutants from the filtration of groundwater. KAF has been shown to remove 97% of arsenic, at least 85% of phosphates and 99% of iron when used in Cambodia, and biochar filters are able to remove up 54% of herbicides when pyrolysed at 625°C and up to 100% of herbicides if pyrolysed at temperatures of 900°C. These promising results lead us to believe that we will be able to find a method of filtration that is able to produce safer drinking water for the Cambodian rural areas.

Keywords

arsenic, pesticides, biochar, filter, Cambodia

Introduction

Only 16 per cent of rural Cambodians have access to adequate sanitation and 65 percent to safe water. [1] This has lead to the preventable death of thousands of young children from diarrhoea and water-borne diseases as well as a plethora of health problems for the older generations. [2] In recent years, attempts have been made at resolving this problem through the use of biosand filters. Though having a high efficacy at clearing out micro-organisms and salts, it was still unable to efficiently resolve a key issue of removing heavy metal ions and organic pesticides which are toxic to the body. [3] Heavy metal ions are a major source of concern for groundwater contamination. They stem from sediments carried by rain or river water accumulating over thousands of years. The main threats to human health from heavy metals are associated with exposure to iron, lead, cadmium, mercury and arsenic. One of the most widespread contaminants is arsenic. In Cambodia, it is estimated that between 75,000 and 150,000 people are consuming arsenic contaminated drinking water for at least part of the year (as of 2008). [4] Symptoms of chronic arsenic poisoning include hyperpigmentation, depigmentation, keratosis, skin cancer, internal cancer and even death.

There have been water filters developed to combat this concern. The Kanshan Arsenic Filter (KAF) has proved to be effective at filtering arsenic in groundwater to less than 50µg/L the acceptable standard in developing countries like Cambodia [5]. However, a KAF requires specific conditions to effectively filter out the Arsenic in the water, and it is unable to lower Arsenic concentrations to safe drinking levels if the inlet concentration is above 100µg/L. In addition, it is unable to filter out other pollutants from the water such as mercury, cadmium and uranium which gets into the soil through pesticides. On the other hand, the use of charcoal as a filtration system has been shown to filter pesticides from drinking water [6]. More specifically, the use of biochar, charcoal made from biomass, is not only cheap and abundant but works as a good absorbent of pollutants. The porosity; large surface area; presence of anions which works as a cation-exchanger and classical graphite structure which enables the carbon to connect with neighbouring atoms or atoms from foreign molecules to establish linear or cyclical bonding makes biochar a stable and effective compound for water filtration [7]. However, biochar is unable to effectively filter arsenic from groundwater.

Hence, the aim of the experiment is to combine both the KAF and the use of biochar into an efficient and cost-effective water filtration system for Cambodia that is able to filter out both microorganisms as well as heavy metal ions - major types of health-impacting pollutants specific to Cambodia.

Methods

A first prototype of the KAF was made according to using readily available materials. A 1.5 litre PEP bottle whose base was slice out was used as the filter container. The bottle cap was then perforated with holes of roughly 1mm radius and 3mm apart and an additional layer of cotton wool was added to the inner layer of the bottle cap to prevent the fine sand from passing through the filter into the drinking water. The layers of the filters were added in the following order and composition- 4cm in height of gravel, 4cm in height coarse sand, 8cm in height of fine sand, 3cm of air layer. A perforated small plastic cup that was used to hold the rusted nails was then tied to the top of the filter container using two strings.

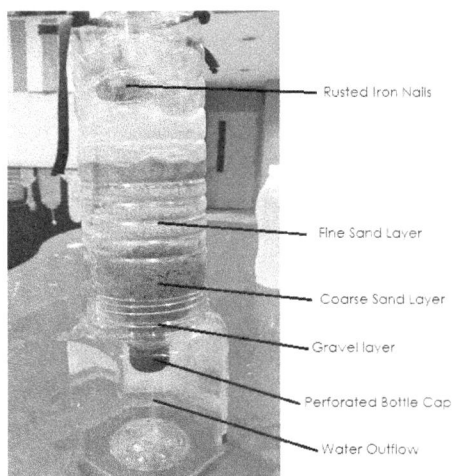

Figure 1: Prototype of KAF

The biochar-KAF is built by incorporating a biochar layer within the KAF layers as shown in Figure 2. The biochar was obtained through the pyrolysis of rice husks which are a waste product generated by the agricultural processes in Cambodia.

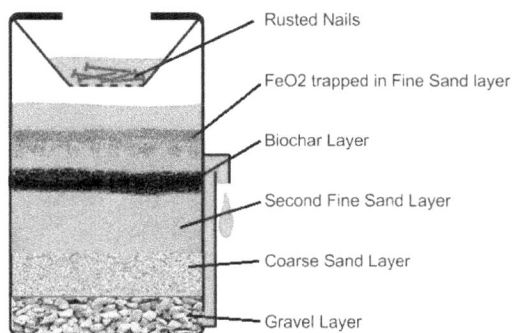

Figure 2: Biochar-KAF Model

Results & Discussion

Table 1: Arsenic Content in Cambodian Soil

Provincial Summary of Arsenic in Cambodia

Province	Tested Wells	Average As (ppb)	% Wells >=50ppb	% Wells >=250ppb	Estimated Impacted Pop.
Kampong Cham	7775	52.5	36.5	6.8	25,552
Kampong Chhnang	1083	11.0	8.1	0.7	3,932
Kampong Thom	1955	10.9	6.6	0.1	2,358
Kandal	17709	94.7	35.9	17.4	69,593
Kracheh	1010	16.3	13.8	1.3	1,144
Prey Veng	10648	32.0	16.8	5.2	40,922

Analysis against current filters

KAF Filters

A review of the KAFs implemented in Cambodia has shown encouraging results. Even with the high levels of arsenic and phosphates present in groundwater, the design used with pre-rusted nails had a 97% removal rate for arsenic. Also, no correlation has been found between flow rate and the amount of arsenic removed, suggesting that this filter does not need a constant supply of water in order to function efficiently. The amount of arsenic also did not increase over the 30 week period for which the filter was tested. There is also almost complete removal of iron at 99%, with phosphate removal between 85% and 88%. E. coli has been shown to be completely removed from ground water after the initial 2-3 weeks.

Biochar

Biochar filters using other materials have been tested in previous studies for the removal of herbicides. Notably, pine pellets have shown better efficiency than the traditional mixture of pine, bamboo, eucalyptus and longan, with almost 100% removal when pyrolysed at 900°C. At lower temperatures of about 625°C, 54% removal is achievable. We believe that rice husks would be able to produce similar yields.

Conclusion

In conclusion , the biochar augmented Kanchan filter has been shown to have improved functionalities in terms of removing hazardous particles such as organopesticides which conventional filters are unable to deal with effectively, we seek to further test and improve this design with NGOs on site in Cambodian rural communities.

Future Work

We seek to do further analysis on the quality of water produced by the biochar-KAF such as the specific arsenic or pesticide content. We also seek to cooperate with non-governmental organizations in Cambodia rural communities to bring this filter to the locals there.

Acknowledgements

We would like to thank Ms Madeline Lim-Chen, Geography Department, teacher and Internal mentor, NUS High School for her continuous support. Dr Chiam Sher-Yi, Research Head, NUS High School for supporting the project. Last but not least, Prof Ng Yong How, Department of Civil and Environmental Engineering for allowing us to use the biochar pyrolysis chamber.

References

[1] "Lack of Adequate Sanitation Triggers Child Health Concerns in Cambodia."UNICEF. Web. 06 May 2014. <http://www.unicef.org/infobycountry/cambodia_39558.html>.

[2] "Water-related Diseases." WHO. Web. 06 May 2014. <http://www.who.int/water_sanitation_health/diseases/cyanobacteria/en/>.

[3] Singh, Reena, Neetu Gautam, Anurag Mishra, and Rajiv Gupta. "Heavy Metals and Living Systems: An Overview." Indian Journal of Pharmacology 43.3 (2011): 246. Print.

[4] "Resource Development International - Cambodia; Ground Water Arsenic Testing." Resource Development International - Cambodia; Ground Water Arsenic Testing. Web. 06 May 2014. <http://www.rdic.org/ground-water-arsenic-in-cambodia.php>.

[5] C. M. Espinoza, M. K. O'Donnell, Evaluation of the KanchanTM Arsenic Filter Under Various Water Quality Conditions of the Nawalparasi District, Nepal, 2011

[6] An Ancient Filtration Material Removes Pesticides from Drinking Water."Engineering for Change. Web. 06 May 2014. <https://www.engineeringforchange.org/news/2012/12/13/an_ancient_filtration_material_removes_pesticides_from_drinking_water.html>.

[7] Carbon-Terra - Waste Water Treatment." Carbon-Terra - Waste Water Treatment. Web. 06 May 2014. <http://www.carbon-terra.eu/en/biochar/application/Waste_water_treatment>.

An investigation on *Ipomoea Aquatica*'s efficacy as a bio-remediator and its market feasibility in Singapore

Priyadharshini Santhanakrishnan, Madhumita Narayanan

NUS High School of Mathematics and Science, Singapore, nhsczm@nus.edu.sg

Abstract

Ipomoea aquatica also known as Kang kong is a fast thriving aquatic plant that is native to the Southeast Asia. Optimum yields can be obtained in lowland humid tropics under reasonably high, yet stable temperatures of between 25–30°C. This makes the plant to be ideal to be grown in Singapore. This project proposes that freshwater rivers can be efficiently used to grow *Ipomoea aquatica* that can contribute greatly to the food demand in Singapore. It is efficient in preventing eutrophication by reducing the amounts of nitrogen, phosphorus and can reduce the water pollution caused by the proliferation of cyanobacteria and algae. After cultivation, the plants can be harvested in about 2-3 weeks to be transported to local markets. In this process, the water bodies will also be cleaned and the carbon dioxide content in the air can also be reduced. We have tried this in a smaller scale in our school's Eco-pond and the plants have shown promising results in their growth and also bio-remediation.

Keywords:

Ipomoea aquatica, kang kong, bio-remediation, hydroponics.

Introduction

Ipomoea Aquatica also known as Kang kong is a fast thriving aquatic plant [1] that is native to the Southeast Asia. Optimum yields can be obtained in lowland humid tropics under reasonably high, yet stable temperatures of between 25–30°C. This makes the plant to be ideal to be grown in Singapore. *Ipomoea Aquatica* possesses high nutritional value and has a high demand in the food industry in Singapore. However, 90% of the food products consumed in Singapore are imported from our neighbouring countries such as Malaysia, China and Thailand and the agricultural sector in Singapore only makes approximately 0.1% of our Gross Domestic Product (GDP), and it has been constantly decreasing over the last few decades [2]. This high amount of imported goods is usually attributed to the lack of land and natural resources in Singapore. However, we have easily ignored the fact that other than farms, we have other land resources that could be used for agricultural purposes. Our freshwater rivers can be efficiently used to grow *Ipomoea Aquatica* that can contribute greatly to the food demand in Singapore. Furthermore, past researches have shown that *Ipomoea*

aquatica can be efficient in preventing eutrophication by reducing the amounts of nitrogen, phosphorus and other important minerals from the waters that they grow in [3]. This reduces the water pollution caused by the proliferation of cyanobacteria and algae. Singapore's water bodies being potential target for frequent water pollution, growing *Ipomoea Aquatica* can make a great difference in pollution level of the water. Moreover, these plants have shown promising results in the removal of heavy metal ions in the water bodies that they are cultivated in. With heavy metal poisoning as an issue for concern in many countries, this method will be cost efficient compared to the current methods and also environmentally-friendly. Thus, in this project, we aim to grow *Ipomoea Aquatica* in the freshwater bodies of Singapore to test its efficacy as a bio-remediator and also investigate its market feasibility in Singapore's food industry. By reducing the percentage of food products imported from other countries, the carbon footprint can be reduced greatly and it also has the potential to increase the agricultural sector's contribution to our GDP.

The cultivation of *Ipomoea aquatica* in freshwater bodies is time and cost-efficient. It includes 2 stages. The first stage is growing them on seedling trays with substrates such as soil or sponge. This makes use of simple hydroponic techniques. The plant would require the substrate until germination occurs. After germination occurs, the plant then can be transferred to the water bodies for development. They do not require any substrate and can be left to float freely on the surface of the water. However, the place should be cordoned off using a net to prevent the aquatic animals from feeding on them and also plant migration. The plants can be harvested in about 2-3 weeks to be transported to local markets. In this process, the water bodies will also be cleaned and the carbon dioxide content in the air can also be reduced. We have tried this in a smaller scale in our school's Eco-pond and the plants have shown promising results in their growth and also bio-remediation.

Past Projects on similar issue and review about them

Many past research projects have investigated the efficacy of Ipomoea aquatica, more commonly known as kang kong or water spinach, in removing mineral ions such as nitrogen and phosphorus in water, and hence reducing the problem of eutrophication, and it has been found that Ipomoea Aquatica is able to remove minerals and also heavy metal ions such as lead and cadmium from water bodies.

According to data obtained from a past research project, about 41.5 to 75.5% of nitrogen and phosphorus from eutrophic water can be removed by *Ipomoea aquatica*.

Significance and Application of Project

The increasing population in Singapore has sparked a lot of discussions among the citizens about the possible increases in demands in transportation services and jobs. However, it is astonishing to know that not many people worried about the increase in demand for basic necessities like food and the possible rise in cost that occur. Over the years, the per capita consumption of food commodities has been on the rise, especially for vegetables. Thus, the rise for the need of food is inevitable. When the demand increases, the percentage of food imported will also increase and thus the increase in carbon footprint involved in the transportation of the plants from other countries. Thus, alternative methods with low carbon footprint are becoming increasingly essential. The current methods also increase the greenhouse emissions due to the high-tech farm machinery that requires high amounts of energy and the transportation process that uses a lot of energy. Furthermore, the process of growing the plants itself can cause high emissions of carbon dioxide into the atmosphere as it involves land clearing and deforestation in most of the cases. Moreover, water pollution is of great concern in Singapore, because of the limited water resources that is present. Hence, it is important that water is treated and conserved in the most effective way. Since Ipomoea aquatica is cheap and easily available, the use of *Ipomoea aquatica* in treating water in Singapore's water bodies would be a cost-efficient and time-efficient method of water treatment that would also not require much manpower to be carried out. It greatly decreases the cost for water purification in our rivers. Overall, implementing this in Singapore's water bodies would allow an efficient method of water treatment as well as a sustainable food source. This project can also be carried out in small scale in schools as part of their environmental projects as compared to other plants; this is easier to manage and thrives better. Other than in Singapore, this project can also be implemented in other developing countries where lead poisoning is a great concern to the health of the society. Many generations of citizens have been affected greatly due to the heavy metal poisoning. Although, with technological growth, there are many ways to eradicate this problem, most of the ways to do not reach the communities there as the literacy level there is low and many people are not even aware of the causes of the problems that they are facing. Thus, if, this project can be implemented there, the problem of lead poisoning can be reduced significantly as past research have shown that *Ipomoea Aquatica* can be a great way to reduce the lead content in water. Moreover, after using the plant for water treatment, the amount of lead and cadmium present in the plant's system was very little compared to the permissible levels of consumption, hence the plant can still be consumed even after using it for bio-remediation and the treatment of water. This will also put an end to the hunger problems in the countries.

Conclusion

As established earlier, *Ipomoea aquatica* is a very suitable as a source of bio remediation as it is cheap, affordable and an easily accessible form of water treatment. Therefore, it can be used in future as an alternative source of water treatment, and this would save resources spent by country on water treatment. This is extremely significant as Singapore is a country with limited water resources, and currently there are other cleaning methods, but these are expensive and excessive money is spent on it.

Moreover, it is also feasible as a sustainable food source in Singapore, because it thrives well, it is easily cultivatable, and hence it is easily available for everyone. Currently, there is a high amount of import of food from other countries, because the Singapore government's food import policy is to guarantee a steady and sufficient supply of healthy and quality foods from a broad number of countries. However, this increases carbon footprint, as emissions from import/export is a major cause of increasing carbon footprint.

In conclusion, growing *Ipomoea aquatica* is most or all of Singapore's water bodies would result in a cost-efficient and feasible method of water treatment as well as a source of sustainable food supply, and this would be extremely beneficial to the country. This project also has a potential to raise awareness among the youth about the problem of high carbon footprint due to import of food products. If this project gives expected results, it can be done in a larger way, developed into a community project and extended to more places in Singapore and be led by people living that particular residential area. This will further decrease the cost and increase the efficiency of the project. While we are finding for new and extraordinary solutions to the problems of global warming, climate change and the greenhouse effect, we tend to forget the easy solutions that can solve most of the problems at one goal. Many people think that reducing carbon footprint would mean reducing their economical activities which may affect their economy greatly. However, they forget that there are many other ways that would help to do the same activity with a lower carbon footprint.

References

[1] http://www.survivalfoodplants.com/kang-kong-ipomoea-aquatica/

[2] http://www.tradingeconomics.com/singapore/agriculture-value-added-percent-of-gdp-wb-data.html

[3] http://www.sciencedirect.com/science/article/pii/S0378377408000176

[4] http://www.sciencedirect.com/science/article/pii/S0043135407002606

[5] http://www.savefoodcutwaste.com/tips-for-individuals/grow-and-buy/grow-your-own-food/

[6] http://link.springer.com/article/10.1023%2FA%3A1010727325662?LI=true

[7] http://www.tradingeconomics.com/singapore/food-imports-percent-of-merchandise-imports-wb-data.html

[8] http://www.mongabay.com/history/singapore/singapore-physical_setting.html

[9] http://environmentlanka.com/blog/2008/bioremediation-for-water-purification-a-case-study-at-st-coombs-lake-talawakelle/

[10] http://www.lowcarbonsg.com/tag/food-import/

[11] http://rmbr.nus.edu.sg/dna/organisms/details/810

[12] http://www.adaptationlearning.net/singapore/profile

[13] http://www.ava.gov.sg/AVA/Templates/AVA-GenericContentTemplate.aspx?NRMODE=Published&NRNODEGUID=%7b5F284016-6F16-4374-91A6-12C9D9E0734E%7d&NRORIGINALURL=%2fAgricultureFisheriesSector%2fFarmingInSingapore%2fHorticulture%2f&NRCACHEHINT=Guest#3

Appendix

Suppl. Fig. 1

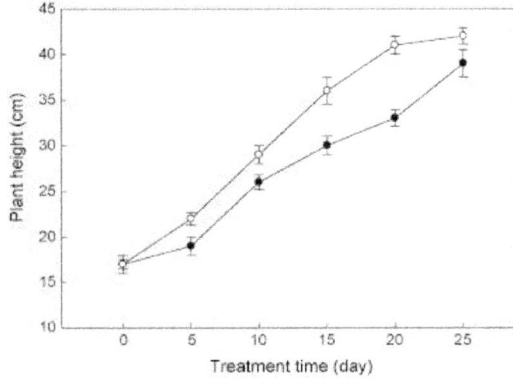

Source: http://ars.els-cdn.com/content/image/1-s2.0-
S0378377408000176-gr4.jpg

Suppl. Fig. 3

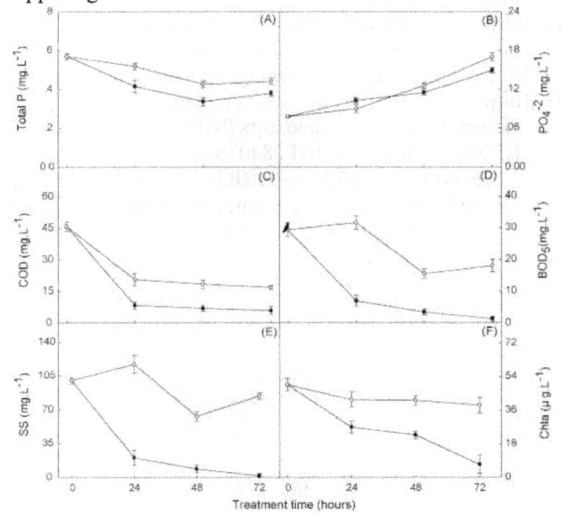

Source: http://ars.els-cdn.com/content/image/1-s2.0-
S0378377408000176-gr2.jpg

Suppl. Fig. 2 Agriculture in Singapore's GDP

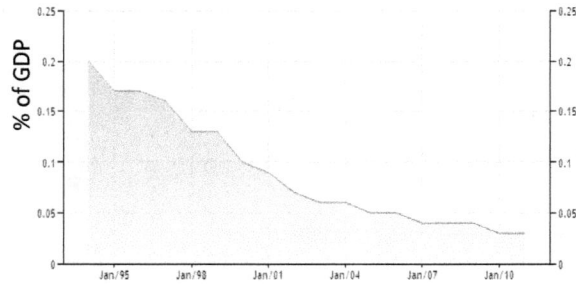

Source:
http://www.tradingeconomics.com/charts/singapore-
agriculture-value-added-percent-of-gdp-wb-
data.png?s=%2fsingapore%2fagriculture-value-added-
percent-of-gdp-wb-
data.html&d1=19940101&d2=20130331&type=area

A Study on the Porous Capacity of Pavement Materials to Alleviate Waterlogging

Liya Hu, Yue Pan, Jincheng Luo, Liwei Feng, Yuqi Wang, Chendan Luo

Wuhan Experimental Foreign Languages School, China, 8750591@qq.com

Abstract

Our hometown Wuhan is a big old city which is in the process of modernization. There are many problems in the area of urban infrastructure. Every year during the plum rain season, it suffers from serious waterlogging caused by torrential rain. One of the major contributors of waterlogging is found to be road hardening. Our aim is to alleviate urban waterlogging by means of water seepage. Depending on the purpose of penetration, seepage engineering are concentrated into three cases. First, our main purpose is to control runoffs. The second is to improve the permeability coefficient of the materials, but without adjusting rainwater storage (groundwater recharge) or controlling peak flow as the main target. According to practical circumstances, we combine non-porous materials with seeping facilities in the third case and achieve a certain design standards of filtration system using facilities in proper sizes and forms. By doing so we can achieve effects that the former means can't. Based on the seeping technology we already have. Our research focuses on these ideas, vegetation, rocks and gravels, which are chosen to be our experimental subjects. Suggestions about the pavement material choosing are put forward based on the experiment results and relative discussion. We hope it could be a model for other cities in similar situations.

Keywords

waterlogging, road hardening, pavement materials, natural porous capacity

1 Introduction

Water is life, but it also can be out of control and disturb our daily life. Thanks to be situated at the confluence of the Han River and Yangtze River, Wuhan has rich water resources. Wuhan's climate is humid subtropical with abundant rainfall in summer. In June, the precipitation is 219.9mm (8.657 inches). High-intensity rainfall challenges the urban drainage system greatly. The rainstorm on June 18, 2011 caused waterlogging in 88 streets, almost paralyzing the entire central city traffic system. Even only in the East Lake Scenic Area, 11.3754 million RMB of economic loss was caused in the storm.

The essential reason of the waterlogging in Wuhan is that even though the urban construction grows rapidly, the drainage facilities haven't kept pace with it. The area of urbanicing in Wuhan now expands to 520.3 square kilometres. The ground of the city is mainly covered with cement, pitch and bricks. At the same time, the seepage technology has been widely used in the drainage systems. Draining the water by seeping has become a possible and effective method. The porous capacity of various pavement materials is examined through scientific experiments. Suggestions about the pavement material choosing are put forward based on the experiment results and relative discussion.

2 Lab Experimental Program

2.1 Introduction

Stormwater infiltration system is generally a simple process, including the interception or pre-treatment measures, infiltration facilities and overflow facilities. Penetration facility may be one or more of the combination.

Depending on the purpose of penetration, studies are concentrated into three cases. First, the initial runoff pollution control for the main purpose. The second is to reduce the loss of stormwater, reducing runoff coefficient, increase rainwater infiltration, but without adjusting rainwater storage and peak flow control requirements. The third is based on adjusting of stormwater storage (groundwater recharge) or control peak flow as the main target, to achieve a certain design standards of filtration system.

The design of infiltration systems will be very different under these three cases. In the first case, it generally does not require special hydraulic and flood storage calculations to use natural ground topography and surface plants mainly, i.e. shallow trench or green buffer zone to absorb pollutants in stormwater to purify surface runoff, to ensure smooth drainage overflow and the soil requirements are low. The second case is somewhat similar to first case, it is also to use more permeable ground, requiring some certain soil permeability, some appropriate storage and hydraulic calculations but without strict requirements on the penetration facilities to ensure the smooth run-off and drainage . The third case is different, a criterion is firstly designed based on storm runoff to determine the need for flood storage or peak flow reduction, to determine the local soil infiltration coefficient and meet the design requirements. The second design criteria accords to the site conditions and one or more suitable facilities penetration is selected by hydraulic calculation to determine the infiltration facilities (infiltration area, length of surface flow and storage capacity, etc.) in order to achieve peak flow control. The overflow of storm water runoff beyond standard should be considered.

2.2 Case 1: Determination of surface runoff

In the first case, one kind of existing ground vegetation and two materials are utilized for their surface runoff determination.

A packing box with a length of 15cm, width 10cm, height 6cm is used as our model. A small hole is opened in the left middle at the height of 4cm and onto which an overflow tube is welded. To maintain stormwater flow, the model keeps 15 degrees with the horizontal. The tube is connected to the cylinder tube until the surface water reaches 250ml. Then the amount of the runoff will be known.

2.3 Case 2: Determination of Soil permeability coefficient

For the second case, the widely used road materials: stones ($\Phi 5 \sim 7$cm), fine stones ($\Phi 0.8 \sim 1$cm) and sand, ordinary soil, are studied as experimental subjects.

The determination of indoor soil permeability coefficient has two methods: fixed and variable water level. The fixed water level is chosen in order to calculate the soil permeability easily.

A plastic bottle (height 24cm, diameter 11cm) with the bottom a 2.5cm diameter flow opening replacing the role of the head is made, and the high water permeability sand is filled to the irregular shape of the bottom of bottle, which its impact can be negligible. The flow opening is covered with a cloth as a protection screen, keeping an average total height of 4cm. At the last, the test material is placed above the sand.

For this series of experiments, the equation for the calculating on the permeability (1) and the equation for the calculating on the rain intensity of Wuhan (2) as the following:

$$K = \frac{Q \times L}{A \times H} \quad (1)$$

$$i = \frac{983[1 + 0.65 \log(P)]}{(t+4)^{0.65}} \quad (2)$$

L - Thickness of soil samples

A - Cross-sectional area of soil samples

H - Constant water level

Q - Constant seepage flow

i – Rain intensity

P – The time of the period between reappearing

t – The time of the duration of the rain

2.4 Case 3: Determination of seepage facilities

For the third case, deep and shallow infiltration wells are chosen according to the water table and geographical conditions. The former applies to a large amount of concentrated water with good quality. The latter is more common in urban areas as a scattered infiltration facility, similar to an ordinary inspection well. However, the wall and bottom are made permeable. Thus, the rainwater can penetrate through the wall, bottom to all around the well. The main advantages include less surface occupation, less underground space, and convenience in management. The disadvantage is that the purification capacity is limited with high storm water quality, which requires less suspended solids.

Based on the second experiment, a plastic bottle (height 12cm, diameter 5cm) with 32 holes on the wall is added, its top cut, a small hole drilled at the bottom and a small tube connected to simulate rainwater pipes. It is inserted upside down into the Case 2 experiment model surrounded by buried gravel and sand layer.

The measurement shall identify the presence or absence of water seepage wells by flow difference.

3 Conclusion

The extensive experiments have been conducted to calculate the rain intensity of Wuhan. As shown in Table 1, the permeability of rocks and gravels are better than the other materials. The Eq. (2) indicates that the ground covered with rocks and gravels can afford the rain that comes once ten years. Normal soil can cope with the rain that comes once a half year. Soil with tiny grains cannot cope with the rain most times. Waterlogging easily occurs in a city with grounds covered with this kind of soil. However, rocks and gravels are not good materials for roads paving, but suit for the greens, parking lots and sidewalks.

At last, the new seeping bricks at the parking lots, rocks and gravels on the greens and seeping wells under them can help relieve waterlogging.

References

[1] Che Wu, Li Junqi (2006). *Technology and Management of Rainwater Utilisation.* (in Chinese) Beijing, China: China Architecture and Building Press. pp. 165–166,169-174.

[2] Parkinson, Jonathan, et al (2007). *Urban Stormwater Management in Developing Countries.* (in Chinese) Beijing, China: China Architecture and Building Press. pp. 113–115.

[3] Sun Li, Wang Xinwen (2006). *Drainage Engineering* (in Chinese). Wuhan, China: Wuhan University of Technology Press. pp. 216–219.

[4] Xv Jing (2011). The Chief of Bureau of Water Resources of Wuhan Was Called to Account for the Urban Inundation. (in Chinese) http://finance.sina.com.cn/roll/20110722/16191019348 7.shtml

Table 1 Data of the Experiments

No.	Material	A(m^2)	H(m)	L(m)	W(m^3)	T(s)	Q(m^3/s)	K(m/s)
1	Rock with Sands	1.364*10⁻²	0.09	0.08	$2.3*10^{-4}$	10.9	$2.11*10^{-5}$	$1.51*10^{-3}$
2					$2.5*10^{-4}$	13.6	$1.89*10^{-5}$	
3	Rocks		0.09	0.08	$2.3*10^{-4}$	5.4	$4.33*10^{-5}$	$3.17*10^{-3}$
4					$2.7*10^{-4}$	7.4	$3.65*10^{-5}$	
5					$2.25*10^{-4}$	5.4	$4.17*10^{-5}$	
6	Soil for Gardening		0.07	0.06	$1.85*10^{-4}$	20.6	$0.65*10^{-5}$	$0.50*10^{-3}$
7					$2.1*10^{-4}$	39.8		
8	Gravels	2.32*10⁻²	0.07	0.05	$1.3*10^{-4}$	1.7	$7.65*10^{-5}$	$4.93*10^{-3}$
9			0.04	0.03	$1.3*10^{-4}$	1.9	$6.84*10^{-5}$	
10	Mud	1.364*10⁻²	0.08	0.07	$2.3*10^{-4}$	5.1	$4.51*10^{-5}$	$3.27*10^{-3}$
11			0.07	0.06	$2.0*10^{-4}$	5.4	$3.70*10^{-5}$	
12					$1.5*10^{-4}$	3.2	$4.69*10^{-5}$	
13	Wet Soil	2.32*10⁻²	0.09	0.08	$1.3*10^{-4}$	1.6	$8.13*10^{-5}$	$1.46*10^{-3}$
14					$1.9*10^{-4}$	4.4	$4.32*10^{-5}$	
15					$2.1*10^{-4}$	7.9	$2.66*10^{-5}$	

Desalination today, water tomorrow: a study in reverse osmosis membrane technology

James Ward, assisted by Grant Gao

Scotch College, Australia, Dina.Poutakidis@scotch.vic.edu.au, Michele.Linossier@scotch.vic.edu.au

Abstract

In 2025, the United Nations estimates that 1.8 billion people will live in areas affected by water scarcity as a result of increased use, exponential population growth, drought and increasingly limited accessibility to water. Fresh water is a precious commodity and seawater is plentiful. Desalination presents a viable solution to the predicted global water crisis. The desalination process involves removing dissolved salts from seawater to obtain fresh, drinking water. Reverse Osmosis (RO), which uses semi - permeable membranes and high pressure pumps, is the main method of desalinating water in Australia. The key to desalination is the RO membranes themselves, hence their importance. The process has not yet been perfected - this technology makes the construction of desalination facilities costly and the operation of them - energy intensive. Further development in the materials that are used to construct these membranes must also be contemplated so efficiency and use of energy can be improved and thus, operating costs reduced. Our investigation will predominantly be based on the PA RO membrane and the correlation between membrane structure and performance. We will draw on existing knowledge and explore new developments toward the next generation of RO membranes, which is essential if we are to cope with the 21st century challenges of adequate fresh water supply.

Keywords

Desalination; Reverse osmosis; Polyamide membranes; Cellulose acetate membranes; Australia

1. Introduction

According to the United Nations, water use has grown at more than twice the rate of population increase in the last century- a concerning sign [1]. As stated by international water expert Professor Frank Rijsberman, a person's diet - not the amount of time they spend in the shower - is the major determining factor of water consumption per head. It takes 70 times the amount of water we use in the home to produce the food we consume. Every day, the world's industries use on average four times as much water as people use in their homes [2]. Likewise, more than half of all hospital beds in Sub-Saharan Africa are occupied by patients suffering waterborne illnesses such as cholera. Even in the United States, waterborne diseases cost the country an estimated $50 million every year in medical costs. Access to clean drinking water would prevent 3 million deaths a year [3]. This colourless, odourless and tasteless liquid is a molecule essential for life.

Water is vital to all living things. Yet, of all the world's total water, 97.5 per cent is saline water, while only 2.5 per cent is fresh water, leaving less than one per cent available to living things [2]. The effects of increased use, exponential population growth, droughts and increasingly limited accessibility to water are increasing global water scarcity [4]. Desalination offers a possible resolution to the global issue of an inadequate supply of fresh water. While fresh water is essentially becoming a precious commodity, seawater remains abundant.

Desalination is the process of removing salts from seawater and groundwater (brackish water) to produce potable drinking water.

There are different methods used around the world to desalinate seawater, which can be classified as either thermal based or membrane based technologies. Thermal desalination technologies rely on evaporation processes, for instance, distillation processes (multi-stage, flash and vapour compression), whereas the membrane technology is dominated by reverse osmosis (RO) and nanofiltration (NF) [5-7]. The choice of technology is based upon feed salinity and other aspects, namely labour, energy, cost and land availability. Around the world, RO technology is predominantly utilized and considered one of the most energy efficient desalination technologies [8]. Whereas, in Middle Eastern countries, for example, where fossil fuel supplies are prevalent, the distillation technique is preferred.

RO is a pressure-driven filtration process through "tight" membrane pores. This requires high pressure to force water molecules through semi-permeable membranes to reject the salt in the concentrate stream. For seawater desalination, the operating pressures range from 55 to 70 bar and there is a limit to the amount of fresh water that can be recovered from the feed without causing membrane fouling. Seawater RO plants have recovery rates from 25 to 45 per cent [5, 8].

Nowadays, polymeric, thin-filmed composite membranes are the most widely used, such as polyamide and cellulose acetate blend membranes. Each material possesses different properties which dictate the surface charge [9].

In Australia, Victoria's own Wonthaggi desalination plant is an example of RO in application, potentially producing up to 550 mega litres of clean water per day and supplying up to 33 per cent of Melbourne's annual water demand. The Wonthaggi desalination plant was originally commissioned in the late 2000's, at a time when Australia was facing heavy water restrictions due to a severe drought and rapid population growth. The desalination plant complements the existing capabilities of dams around Victoria while also providing a fall-back option in times of severe water shortage. Despite the high operation expenditure associated with such plants, the benefits of desalination outweigh the costs. Governments, along with organisations such as the World Health Organization (WHO) and United Nations International Children's Fund (UNICEF), would greatly benefit from RO desalination plants to provide a clean and reliable water source.

This study focuses on the review of the overall process of desalination which operates using RO membranes based technology, adopted in the local desalination plant at Wonthaggi. Furthermore, preliminary experiments were also conducted on the salt-removal step. Direct measurements of flux and salt rejections were observed using both polyamide (PA) and cellulose acetate (CA) membranes which provide an understanding on the salt-rejection mechanism of using TFC membranes.

2. Literature Review

The important steps in RO desalination plants are elaborated as follows:

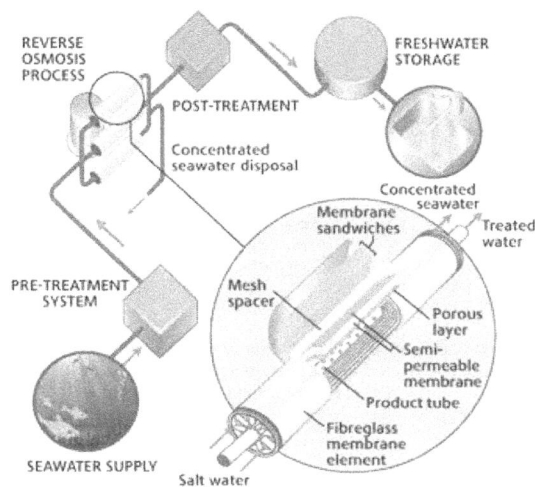

Figure 1: An overview of the seawater RO process [10].

2.1 Seawater Intake

Firstly, the water is drawn in from the ocean from the seawater intake. At Wonthaggi desalination plant, the seawater intake tunnel reaches 1.2 km from the shoreline. This 4 m diameter tunnel lies 20m under the sea bed at the intake point and 40 m below the ground at sea level. The seawater intake and outlet points of desalination plants are designed to have as little effect on local marine life from being drawn into the desalination system. The location of the outlet point also needs to be considered - it is vastly preferable that the outlet is located near a fast moving point of the adjacent body of water, such as a strong current, so that the brine, which contains a significantly higher concentration of salt than that of normal seawater, can be quickly diluted with seawater of a lower salt concentration.

2.2 Filtration Process

The purpose of this filtration process is to sift out large pieces of organic matter before the pre-treatment process commences. At Wonthaggi, the seawater flows by gravity from the sea to the lower chambers of the initial pumping system. From here, the seawater is filtered with a 3 mm screen and then pressurised. The pre-treatment and RO processes both require increased water pressure to achieve their optimal efficiency.

The pre-treatment process exists solely to remove substances that would otherwise reduce the effect of the RO membranes and/or cause fouling of the membranes. The pre-treatment process varies greatly depending on the source of water and water quality at the intake point, along with the composition of the feedwater. In other words, different potential causes of fouling have differing appropriate pre-treatments which apply in their specific circumstances.

Table 1: Different potential causes of fouling that apply across a broad range of seawater RO desalination plants [11].

Fouling	Cause	Appropriate pre-treatment
Biological fouling	Bacteria, microorganisms, viruses, protozoan	Chlorination
Particle fouling	Sand, clay (turbidity, suspended solids)	Filtration
Colloidal fouling	Organic and inorganic complexes, colloidal particles, algae	Coagulation + Filtration Optional: Flocculation / sedimentation
Organic fouling	Natural organic matter	Coagulation + Filtration + Activated carbon adsorption Coagulation+ Ultrafiltration
Mineral fouling	Calcium, Magnesium Barium or Strontium sulphates and carbonates	Antiscalant dosing, Acidification
Oxidant fouling	Chlorine, Ozone, $KMnO_4$	Oxidant scavenger dosing: Sodium bisulfite, Granulated Activated Carbon

At Wonthaggi, before the commencement of fine filtration, ferric sulphate, chlorine and sulphuric acid are added. The sulphuric acid serves as pH control, while the ferric sulphate serves as a coagulant - a substance that acts as a catalyst for a process of adhesion whereby particles in the water group to form larger-size clusters which can be easily filtered.

The chlorine acts against biological organisms in the feed water.

The primary component of the pre-treatment system involves a series of tanks which house filtration membranes. This filter additionally consists of 10 cm of gravel, 40 cm of sand and 50 cm of anthracite, which further act as a filtration system for any possible sediment.

The pre-treatment process is finalised by forcing the feed water through ultra-fine filters to further protect the RO membranes from any large particles that may have passed through the pre-treatment. In addition, the feed water is de-chlorinated with sodium bisulfite to prevent oxidation on the RO membranes.

2.3 Reverse Osmosis

The key to desalination is the RO membranes themselves, hence their importance. In the RO pressure-driven filtration process, fresh water is produced by separating the dissolved salts using polymeric membranes.

In a normal environment, solute (or ion) spontaneously diffuses from a concentrated system to a more dilute environment due to osmotic pressure gradient. Osmosis is a passive process, and the process stops when the system reaches equilibrium i.e. the concentration gradient driving force, is nullified. In this condition, the system is in minimum energy state, free energy is equal to zero. Conversely, reverse osmosis is the process of using pressure to force a liquid, with a high solvent concentration, through a membrane, thus producing pure water. To counter the osmotic pressure, and remove the clean water from the concentrated system, significant energy in the form of applied pressure is required to "push" the water through the membrane.

All other practises involved in RO desalination relate either directly or indirectly to support and assist the RO process. RO membranes provide the most essential, but complex, function in desalination - the rejection of salt.

The higher the osmotic pressure (that is, the salt concentration in the feed stream), the higher the energy required to drive the separation in the reverse direction. To accomplish the separation process, the semi-permeable, "tight pore or dense" membrane is utilised. The polymeric membrane possesses surface charge due to a polymer functional group. This surface charge enhances the salt rejection through ionic repulsion aside from rejection due to steric hindrance (that is, physical sieving process) [13].

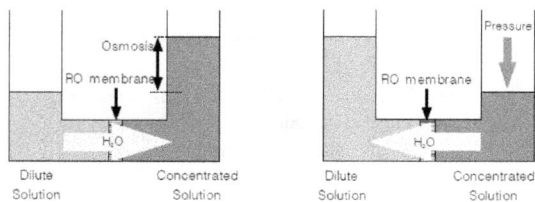

Figure 2: Reverse Osmosis [12]

Pressure required to force the feed water molecules through a given membrane increases when the salt content does likewise.

A RO membrane rejects salt via three methods: size exclusion, charge repulsion and absorption. RO membranes are designed so that they can reject salt ions solely based on their size - water molecules are smaller than salt ions and thus, are able to pass more freely across the semipermeable membrane. Salt ions are also rejected based on their charge; RO membranes are negatively charged; which allows them to repel negatively charged

chloride ions based on charge repulsion. However, as the levels of salinity increase, more ions accumulate on the membrane surface, shielding the actual membrane charge; the effectiveness of charge based repulsion is decreased. The third way that RO membranes reject salt ions is through absorption of the actual chloride ion [13].

Seawater usually has a salt concentration of around 35 grams per litre (the salinity of water found at the point of intake for the Wonthaggi plant drawing from Bass Strait is typically around 37 grams per litre) thus, the driving force required to achieve an efficient flux rate typically ranges from 50 to 60 bar (5000 to 6000 kPa).

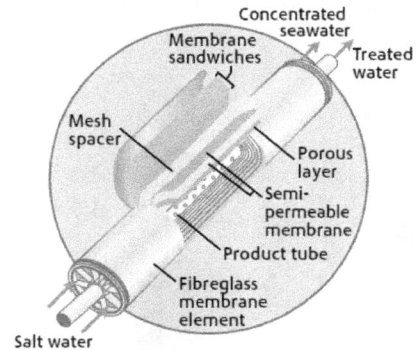

Figure 3: A graphical depiction of a semi-permeable RO membrane used at Wonthaggi and other seawater RO desalination plants around the world [10].

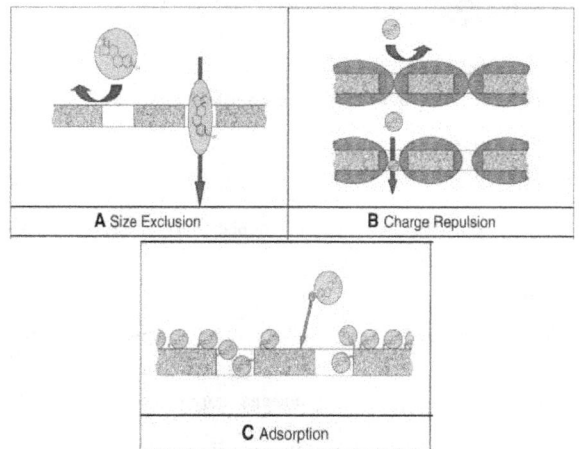

Figure 4: Membrane rejection techniques [13]

In industry, "spiral wound" membrane modules are utilized. This configuration offers high surface area per volume ratio [14]. Inside, the module consists of multiple layers of semi-permeable membranes separated by mesh spacers which set apart each level of membrane, while also generating turbulence which aids the membrane filtration process along with reducing fouling potential. In this membrane configuration, the feed water travels through the mesh space, pressing against the membranes located on either side. Permeate which passes through the membrane is collected in the permeate carrier layer which exists between each layer of membranes, and is pushed, in a spiralling fashion to the central product tubing, as shown in Fig.3.

Figure 5: A number of first stage trains at Wonthaggi desalination plant, Victoria.

The RO process at Wonthaggi effectively involves two stages – the first and second pass. Firstly, large pumps pressurise the feed water to a sufficient rate (around 55 bar), then the feed is forced through large arrays of membranes, known as trains.

Following pre-treatment, the water is pumped to the actual RO process, where it is forced through a series of semi-permeable RO membranes.

The piping used until this point in the desalination process has been made of reinforced fibreglass rather than stainless steel. This is because the fibreglass is not as prone to corrosion from salt as the latter.

The recovery rate (the percentage of fresh water obtained from seawater) of the first pass is only 48 per cent. During this pass the majority of the brine is removed from the process. The permeate (water that passes through) of first pass, is then pumped through to a separate set of membrane trains that make up the second pass. The recovery rate is significantly higher, at around 90 per cent, while the 10 per cent that doesn't pass through is recycled into the first pass feed.

At Wonthaggi, each RO membrane is 50 mm in diameter, 1.2 m long and each of the membrane pores are 5 microns wide. Each individual membrane has a surface area of 41 m^2. The first pass of each of the three sections consists of 9 'trains' of membranes, each containing 218 RO membranes. Only 8 of the nine trains remain in operation at one time, as the remaining train provides an extra slot so that membranes can be flushed out to remove fouling.

2.4 Energy Recovery

The concentrated stream which contains a high amount of salt is known as brine concentrate (retentate). At Wonthaggi along with most other seawater RO desalination plants, this water is used for the energy recovery process that recycles almost 50 per cent of the kinetic energy used to force the feed through each pass. This is achieved through energy recovery units, like those made by Energy Recovery Incorporated, which capture hydraulic energy from the high pressure brine reject

streams and transfer this energy to lower pressure feed water with an efficiency of over 98 per cent. Furthermore, the device also requires no electrical power; therefore, it is more environmentally efficient.

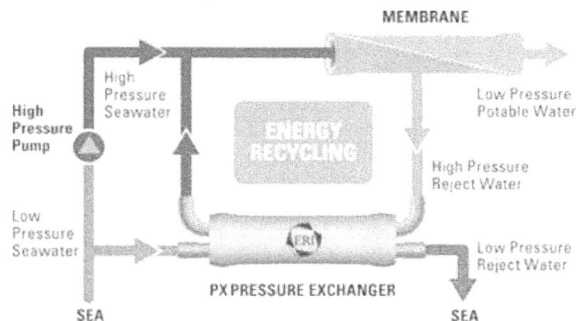

Figure 6: Energy Recovery set-up [15].

The high pressure reject brine turns the only moving part in the design, a rotor that turns up to 1,200 rpm. This rotor transfers the kinetic energy of the high pressure brine to the low pressure seawater feed, complimenting the high pressure pump's own capability and thus, reducing operating costs. Energy recovery units are a particularly notable example of processes developed to make desalination more efficient and thus, less environmentally obtrusive.

2.5 Post-treatment

Following the RO and energy recovery processes, the permeate is re-mineralised. The water that is produced by the desalination process is devoid of many minerals that are usually found in tap water. It is for this reason, that at Wonthaggi, the permeate water is re-mineralised with small quantities of carbon dioxide, calcium hydroxide, hexafluorosilicic acid (for fluoride addition) and sodium hypochlorite.

3.   Method

To understand the flux and salt rejection of RO membrane, experiments were conducted using flat-sheet membrane modules, as illustrated in Fig.7. The membranes utilized were polyamide membranes (DOW Film tech SW30) and cellulose acetate blend membranes (GE Osmonics Flat sheet membrane).

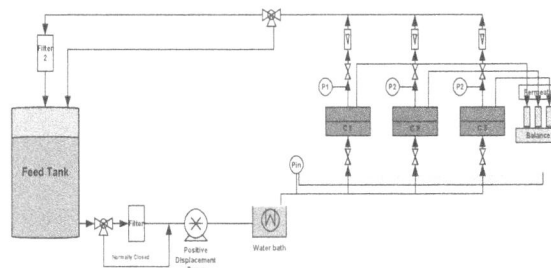

Figure 7: The schematic diagram of the experimental set up [16].

The membranes were mounted with the active surface facing the feed stream. The feed spacer was equipped to enhance turbulence and permeate was drawn from the top of the compartment. The membrane active area for each cell was exactly 0.0042 m^2.

A Hydracell G-03 high pressure pump was utilized to direct the flow through the membrane module, while back

pressure regulators were equipped to adjust the pressure. The flux and rejection were measured at three different pressures 5000, 5500 and 6000 kPa, respectively. The flux was monitored by recording the mass of the permeate for 2- 4 minutes. The sodium chloride, NaCl (ChemSupply, 99 per cent) concentration was measured using a conductivity meter, CRISON basic 30+. A linear calibration curve was constructed to calculate the actual concentration.

The salt concentration tested varied between 2 to 30 g/L. At the highest salt concentration of 30g/L, both polyamide and cellulose acetate membranes were utilized.

4.   Results and Discussion

The rejection rate of an RO membrane is dictated by two main factors: surface charge and size exclusion (that is the pore size of the membrane). Both membranes types are negatively charged so as to repel the chloride ions (Cl$^-$). In low concentration systems, the surface charge repulsion is highly effective. Yet, as the system gets more concentrated, the ions are accumulated on the vicinity of membrane surface which screen the membrane charge, resulting in the decrease of salt rejection.

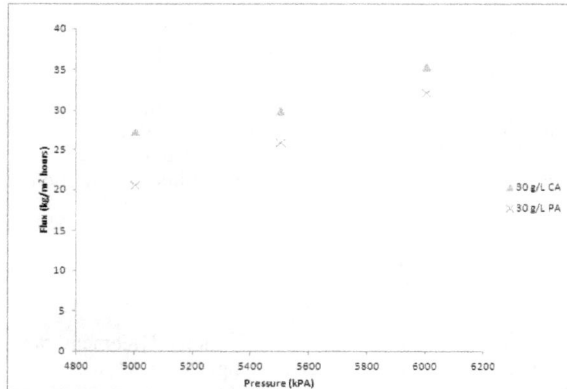

Figure 8: Permeate flux of PA and CA membranes with respect to increasing pressure.

The permeate flux of CA and PA membrane at 30g/L NaCl is depicted in Fig.8. It can be seen that with increasing pressure (driving force) the flux increases in a linear manner. The observation showed that the flux of the PA membrane is marginally lower than flux of the CA membrane. As PA is slightly more hydrophilic than CA, it is expected that the flux for PA is slightly higher than CA. However this discrepancy might be due to error measurement, condition of the membrane and subtle measurement of the permeate flux.

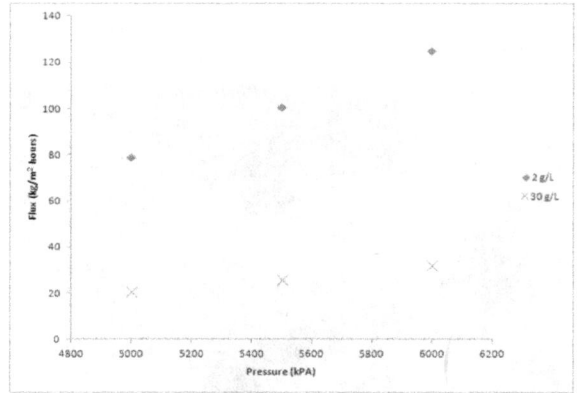

Figure 9: Flux of PA membranes in different salinity levels with respect to increasing pressure.

Fig.9 depicts the linear increase of flux with respect to pressure as previously explained. Additionally, with increasing salt concentration in the feed tank, the flux is significantly reduced. By increasing feed salinity from 2 to 30 g/L, the permeate flux is reduced by a factor of 4 to 6 times. As the salt concentration increases, osmotic pressure also increases, which reduces the overall driving force, resulting in considerable decrease in product flux.

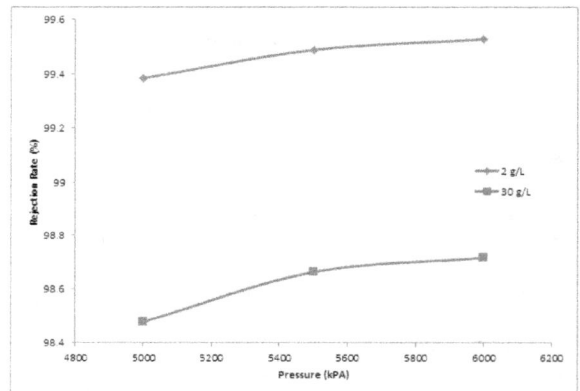

Figure 10: Rejection of PA membranes with respect to increasing pressure and salt concentration.

It is expected that the salt rejection would increase with respect to applied pressure as shown in Fig.10. With increasing driving force (pressure), water molecules travel much faster compared to the larger sodium ions (Na$^+$) and chloride ions. Furthermore, as the membrane itself acts as an electro-repulsion shield for chloride ions (Cl$^-$), they cannot easily pass through the membrane. In the solution, the amount of positive and negative ions has to be equal (electro-neutrality within the solution), thus, as Cl$^-$ are rejected, the paired Na$^+$ ions cannot pass through the membrane.

Additionally, it is also expected that the rejection would significantly fall with increasing salt concentration. As forementioned, with increasing total ions in the solution, more ions are accumulated on the membrane surface which screens the repulsion charge of the membrane.

With higher amount of water pass through the membrane (at higher pressure) compared to the ions leaked through the membrane, the concentration of salt in the permeate decreases, which means increase of salt rejection.

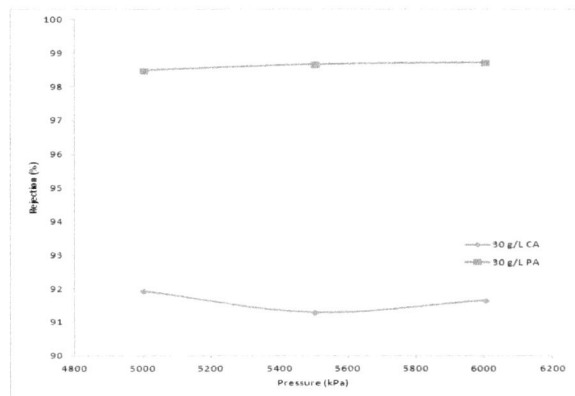

Figure 11: Rejection of PA and CA membranes with respect to increasing pressure.

Rejection of PA membrane is expected to be much higher than CA as featured in Fig.11. PA is more hydrophilic with higher membrane surface charge density which greatly enhances the surface charge repulsion. The behaviour of increasing rejection with increasing applied pressure is still expected, yet this effect is not observed for CA membrane. This might due to the inaccuracy during measurement. Furthermore, as the contribution of the surface charge of the CA membrane is less apparent than in PA membrane, with accumulation of ions on the membrane surface, the membrane charge is highly diminished. In this case, with higher applied pressure, more ions are able to be pushed through the membrane along with the water and hence the expected trend of increasing rejection with increasing pressure is not clearly observed.

4.1 Comparing PA and CA membranes

From the combination of experimental testing and further research, many differentiating factors of PA membranes to CA membranes were revealed.

"Cellulose acetate and aromatic polyamide group of polymers are known as two of best polymer materials for RO application till date" [9].

The first commercially feasible RO membrane was the cellulose acetate membrane - which consisted of an asymmetrically structured thin film acetone-based solution of CA polymer. This membrane type has two layers: a dense surface layer of about 0.1 - 0.2 µm, responsible for the salt rejection property and a 100-200 µm thick, spongy, and porous layer which supports the surface layer mechanically and has high water permeability [12].

Conversely, the thin film composite (TFC) PA membrane is distinguished by higher specific water flux and salt rejection than CA membranes. Additionally, they are stable over a wider pH range and operable at lower pressure than CA membranes [12].

CA membranes have an initial lower purchase cost compared to that of PA membranes. They are also tolerant of quantities of chlorine in the feedwater. From the combination of experimental testing and further research,

it was discovered that the flux generated by CA membranes is higher than that of PA membranes due to the smaller surface area required (Fig.8). However, CA membranes are less stable outside of their small pH operating range of four to eight, which may partially explain the results depicted in Figure 11, wherein the rejection did not appear to increase in a linear fashion. The inferior salt rejection of CA membranes compared to that of PA membranes was also demonstrated by our findings (Fig.11). This is primarily due to the fact the CA membranes are less charged than PA membranes - thus the smoother surface than PA membranes. CA membranes also suffer from compaction effects under pressure. Furthermore, CA membranes have a lower permeability and thus, require higher operating pressures. Additionally, CA membranes are prone to microbiological attack, although this can be easily prevented with the addition of chlorine to the feedwater, which CA membranes boast tolerance to.

Conversely, PA thin film composite membranes offer greater membrane stability and membrane life, along with superior salt and organics rejection as demonstrated in Fig.11. They also boast a wide temperature operating range and a wider pH operating range compared to that of CA membranes. Additionally, PA membranes are less susceptible to microbiological attack than CA membranes. For these reasons, PA thin film composite membranes are currently the most widely used desalination. However, PA membranes have a limited tolerance to chlorine which constrains their life and performance characteristics - they suffer from oxidation at exposure. Moreover, de-chlorination itself can lead to biofouling of the membrane. Hence, the importance of the development of chlorine resistant membranes.

This flaw ultimately leads to stringent pre-treatment requirements. Furthermore, high feed pressures applied in the RO process can potentially damage PA membranes internally - physical compaction of the internal support membrane which creates irreversible internal fouling - a concern that remains for RO membranes to date. As well as this, PA membranes have a higher purchase cost, although in large desalination facilities like that in Wonthaggi, the operating costs of CA membranes combined with their lower efficiency outweigh the initial cost of PA membranes which additionally have a longer operating life [9, 17].

5. Conclusion

The question remains, why are 1.76 billion people faced with a shortage of fresh water when there is a plentiful supply of seawater that can be utilized? In response to this global dilemma, desalination offers a feasible solution to this crisis of water, hence the increasing implementation of desalination plants around the world. Even so, the quest for more energy and cost efficient desalination technologies is paramount.

The correlation between membrane structure and performance is a critical aspect regarding efficiency. Today there are two main types of membranes - the cellulose acetate and its successor the polyamide TFC membrane. Although polyamide membranes have now effectively replaced the original cellulose acetate membranes, a wide area of research is still currently conducted to improve their performance, efficiency, chemical resistance and mechanical strength.

Another technique, namely ion exchange and the absorption process could potentially show better performance. Zeolite crystals, could possibly offer chloride resistance, yet it might be susceptible to biological fouling residue present in the system. This alternative however, could reduce energy costs, in tandem with improving salt salt selectivity, flux and anti-fouling properties [9].

As chlorine can damage the membrane and thus lead to lower salt rejection and poor quality permeate, the development of chlorine resistant membranes is another future process to consider [9]. Taking into account that RO membrane technologies are 'evolving' with the production of a new generation of membranes, the future horizon of desalination looks promising if we are to cope with fresh water shortage.

Acknowledgments

Ms K. Kezia, PhD Student.

Mrs D. Poutakidis, Science Teacher, Scotch College.

Ms M. Linossier, Academic Extension Coordinator, Scotch College.

References

[1] UN - Water Thematic Initiatives (2006), *Coping with water scarcity*, p. 2.

[2] Lukins, N, Elvins, C, Lohmayer, P, Ross, B, Sanders, R, Wilson, G (2011), *Chemistry 1* (4th ed.), Heinemann, Australia.

[3] World Health Organisation (2009), *Water scarcity fact file*, http://www.who.int/features/factfiles/water/en

[4] United Nations (2014), *Water scarcity among critical food security issues in Near East and North Africa – UN*, http://www.un.org/apps/news/story.asp/story.asp?NewsID=47181&Cr=Food+Security&Cr1

[5] Fritzmann, C, Lowenberg, J, Melin, T, Wintgens, T (2007), *State-of-the-art of reverse osmosis desalination*, Desalination, 216, pp. 1-76.

[6] Baker (2000), *Membrane technology and applications* (2nd ed.), Wiley, USA.

[7] Karagiannis, I.C, & Soldatos, P.G (2008), *Water desalination cost literature: review and assessment*, Desalination, 223, pp. 448-456.

[8] Charcosset, C (2009), *A review of membrane processes and renewable energies for desalination*, Desalination, 245, pp. 214-231.

[9] Ghosh, A.K, Bindal, R.C, Prabhakar, S, Tewar, P.K (2011), *Composite polyamide reverse osmosis (RO) membranes – Recent developments and future directions*, Barc Newsletter.

[10] WaterSecure, *Desalination – fresh water from the sea*, http://www.watersecure.com.au/pub/what-we-do/desalination

[11] LennTech, *Desalination pre-treatment*, http://www.lenntech.com/processes/desalination/pretreatment/general/desalination-pretreatment.htm

[12] SAEHAN Industries, Inc., *Introduction to reverse osmosis membrane*, http://www.csmfilter.co.kr/searchfile/file/Tech_manual.pdf

[13] Kezia, K, Lee, J, Kentish, S (2014), *Desalination: reverse osmosis*, Powerpoint Presentation, Membrane Technology Group, University of Melbourne.

[14] Bodalo-Santoyo, A, Gomez, E, Martin. M.F, Montesinos, A.M (2004), *Spiral-wound membrane reverse osmosis and the treatment of industrial effluents*, Desalination, 160, pp. 151-158.

[15] *Energy recovery Inc. forging ahead in efficient desalination*, image, http://www.examiner.com/article/energy-recovery-inc-forging-ahead-efficient-desalination

[16] University of Melbourne (2014), *Melbourne School of Engineering Standard operating procedure cross flow filtration rig.*

[17] Dow Water and Process Solutions (1995), *A comparison of cellulose acetate and FILMTEC FT30 membranes*, Dow Water.

BIOMIMETIC MEMBRANE ON WATER PURIFICATION

[1]Shuning Xu, [1]Cher Ying Foo and [2]Yen Wah Tong

[1]*Raffles Girls' School (Secondary), Singapore, xu_shuning@hotmail.com*
[2]*Department of Chemical and Biomolecular Engineering, National University of Singapore, Singapore*

Abstract

Based on their unique combination of offering high water permeability and high solute rejection aquaporin proteins have attracted considerable interest over the last years as functional building blocks of biomimetic membranes for water desalination and reuse. The purpose of this research paper is to provide an overview of the effect of aquaporins in water purification as well as the relationship between the thickness of the vesicle and the efficiency in water purification in terms of osmotic water permeability. This experiment was conducted to determine the relationship between the thickness of the biomimetic membrane and their efficiency in water purification in terms of permeability. The osmotic water permeability of the vesicles was derived through substitution of the results from the stopped-flow characterization and dynamic light scattering into an equation. Results showed that the thicker vesicle which is made of polymer with a molecular weight of 6000, P3195- MOXZDMSMOXZ, has a higher permeability value as compared to the polymer with a molecular weight of 3500, P3691B-MOXZDMSMOXZ. Thus, the experimental results support our hypothesis that the thicker the vesicle, the higher the permeability value of the vesicle.

Keywords

Biomimetic membrane, water purification, permeability value

1.1 Purpose of investigation

Over the years, desalination is becoming increasingly important for water production in semiarid coastal regions. Synthetic membranes have come a long way in the 50 years since the invention of the cellulose acetate reverse osmosis (RO) desalination membrane by Loeb and Sourirajan [1]. State-of- the-art synthetic membranes at optimal conditions [2] can now desalinate sea water with an energy demand about 15–20% of that used for the early RO membranes. However this is still 1.5 to 2.0 times the minimum energy dictated by thermodynamics [3]. Consequently, there is a continuing quest for membranes with improved performance to provide better separations at even lower energy demand.

Biological membranes have excellent water transport characteristics, with certain membranes able to regulate permeability over a wide range. The permeability of membranes such as those present in the proximal tubules of the human kidney [4], can be increased by insertion of specific water-channel membrane proteins known as Aquaporins (AQPs).

The AQP-rich membrane, when suitably supported, could be used in similar processes with lower energy or membrane area requirements. Aquaporins are attracting considerable interest because they play central physiological roles. Plants possess a high number of aquaporins that have been involved in numerous processes, such as transpiration, root water uptake, seed desiccation or germination, (Johansson et al., 2000). Recently, several research papers suggest that membranes could achieve better selectivity and permeability from biomimicry as well as incorporating aquaporin proteins into membrane.

Also, the permeabilities of AQP-rich membranes are orders of magnitude higher than those observed for unmodified phospholipid membranes [5]. Additionally, some members of the AQP family have excellent solute retention capabilities for small solutes such as urea, glycerol, and glucose, even at high water transport rates [5, 6] These properties result from the unique structure of the water-selective AQPs. These AQPs have six membrane-spanning domains and a unique hourglass structure [7] with conserved charged residues that form a pore that allows both selective water transport and solute rejection. The AQP used in this study was a bacterial aquaporin from Escherichia coli, Aquaporin Z (AqpZ). AqpZ was selected because it can enhance the permeability of lipid vesicles by an order of magnitude while retaining small uncharged solutes.

High permeability and excellent solute retention of small solutes are important for water treatment in critical medical applications such as dialysis [8, 9] because they could lead to reduced equipment size and more efficient energy use (Pontoriero, 2003). A significant improvement in the permeability of solute-rejecting membranes would also be a large step in improving the economics of desalination for drinking water applications. Aquaporins are pore-forming proteins that can be found everywhere in living cells. Under the favourable conditions, they form 'water channels' that are able to exclude ionic species. In a series of simple characterization experiments, Kumar showed the exceptional water permeability of aquaporins and concluded his observations to postulate desalination membranes with vastly improved performance (Kumar et al, 2007) due to AQP's high permeability and high specificity.

The permeability and solute transport characteristics of amphiphilic triblock-polymer vesicles containing the bacterial aquaporins from Escherichia coli, Aquaporin Z (AqpZ) were investigated. Light-scattering measurements on pure polymer vesicles subject to an outwardly directed salt gradient in a stopped-flow apparatus indicated that the polymer vesicles were highly impermeable. However, a large enhancement in water productivity of up to 800 times that of pure polymer was observed when AqpZ was incorporated. The results showed that the activation energy of water transport for the protein-polymer vesicles corresponded to that reported for water-channel-mediated water transport in lipid membranes [10]. The solute reflection coefficients of glucose, glycerol, salt, and urea were also calculated, and indicated that these solutes are completely rejected (Kumar, 2007). It can be inferred that the productivity of AqpZ-incorporated polymer membranes was larger than values for existing salt-rejecting polymeric membranes. Hence, this shows that the incorporation of aquaporin proteins improve the productivity of water purification. This leads to more productive and sustainable water treatment membranes.

The aim of this experiment is to find out if the thickness of vesicles affect the efficiency of water purification in terms of the permeability to water molecules. In this experiment, we investigate the incorporation of the bacterial water-channel protein AqpZ into an ABA triblock copolymer, PMOXA- PDMS-PMOXA. "PDMS" is a hydrophobic block, while "PMOXA" is a hydrophilic block. Thus, the total molecular weight including both hydrophobic PDMS block and hydrophilic PMOXA block for the 2 polymers are 3500 and 6000 respectively, while 4000 and 2500 are the molecular weight for the hydrophobic PDMS block respectively. By using different types of polymers of molecular weight of 3500 and 6000 for the experiment, we are varying the thickness of the vesicle. Our hypothesis is that the thicker the vesicle, the more permeable the vesicle is to water molecules, thus resulting in a more efficient the water purification process. The permeability value of the vesicles is measured by substituting the values obtained from the Dynamic Light Scattering and Stopped-flow into an equation. The higher the permeability of the vesicle, the faster the speed water molecules pass through it, thus resulting in a more efficient water purification process.

1.2   Materials and Methods

Materials

Amphiphilic polymers used for vesicle fabrication: (1) poly(2-methyloxazoline)-b-   poly(dimethylsiloxane)-b-poly(2-methyloxazoline) (PMOXA-PDMS-PMOXA) with molecular weight of 500-2500-500 Da; (2) PMOXA-PDMS-PMOXA with molecular weight of 1000-4000-

Addition of Aquaporin Z (AqpZ) and biobeads

11.62 µl of AqpZ with detergent of concentration of 0.86 was added to the thin-film membrane and set aside for 5 minutes for the membrane to absorb the protein. 2000 µl of deionized water was pipetted into the flask. A magnetic stirrer was added to the solution which then was stirred at room temperature for 12 hours at the speed of 140. 30 mg of biobeads was added to the sample and stirred for 4 hours. This was repeated 5 times, with a new batch biobeads each time.

Extrusion of vesicles

The resultant vesicles suspension was then extruded 10 times through a filter unit of 0.22 µm to ensure uniform vesicle sizes, which will result in a narrow size distribution of the vesicles.

Stopped-flow Characterization

The permeability of the polymer vesicles was characterized by detecting the light scattering of the preparations in a stopped-flow apparatus. The suspension of aquaporin containing vesicles with initial diameters around 220 nm

was rapidly mixed with the same volume of sucrose buffer, a hyperosmolar solution. The resulting transmembrane osmotic gradient will generate water efflux, which resulted in vesicle shrinkage [11]. The consequent reduction in vesicle volume can be measured as an increase in the intensity of scattered light. The vesicle size changes were monitored and recorded in the form of an increasing signal in the light scattering analysis. The initial rise of the signal curve was fitted to equation (1) below.

$$Y = A \exp(-kt) \quad ----- (1)$$

where Y is the signal intensity, A is the negative constant, k is the initial rate constant (s-1), and t is the recording time.

The osmotic water permeability was calculated using the equation (2) below.

$$P_f = \frac{k}{(S/V_0)V_w \Delta_{osm}} \quad ----- (2)$$

where Pf is the osmotic water permeability (m/s), S is the vesicle surface area (m2), V0 is the initial vesicle volume (m3), Vw is the partial molar volume of water (0.018 L/mol), and Δ_{osm} is the osmolarity difference between the intravesicular and extra-vesicular aqueous solution that drives the shrinkage of the vesicles (osmol/L).The rate constant k of the normalized light intensity increase indicates the rate constant of water efflux, which is proportional to the water permeability coefficient. The light intensity increases exponentially as a function of k with time. The k values can then be used to calculate osmotic permeability.

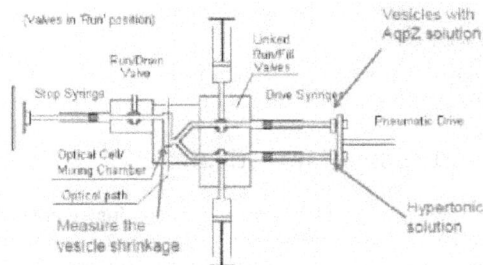

Figure 1: Stopped-flow apparatus

Dynamic Light Scattering

Dynamic Light Scattering (DLS) measures mainly the diffusion coefficient of particles or macromolecules. When monochromatic light is shone through the vesicle solution, it hits the moving particle and scatters off in all directions, resulting in change of wavelength. Before DLS measurements, the samples must be filtered to remove foreign particles so that it will not affect the trend of the graphs. Values from the DLS results which indicate vesicle sizes distribution that will be substituted into the equation to calculate osmotic permeability. If the graph shows 2 or more raised loops, it means that detergent is still present in the solution and the results are inaccurate. Measuring the intensity fluctuations can yield important information about the particles, including the 'hydrodynamic diameter' of the suspended particles, which is an essential piece of information needed to calculate the permeability of the vesicles.

1.3   Results of the experiment

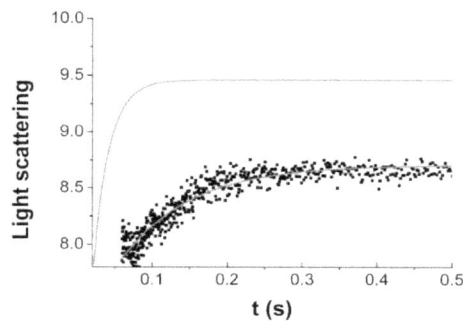

Figure 2.2: Water permeability of the vesicles determined by a stopped-flow apparatus. The increase in the light scattering signal represents a reduction in vesicle size due to water efflux from the vesicle to the solution. The black curve represent light scattering signal for the blank vesicles which are fabricated from PMOXA-PDMS-PMOXA (500-2500-500). The yellow curve represent light scattering signal for the aquaporin incorporated vesicles AqpZ-PMOXA-PDMS-PMOXA (500-2500-500). The data are exponentially fitted by equation (1). The solid lines represent the fitting curves.

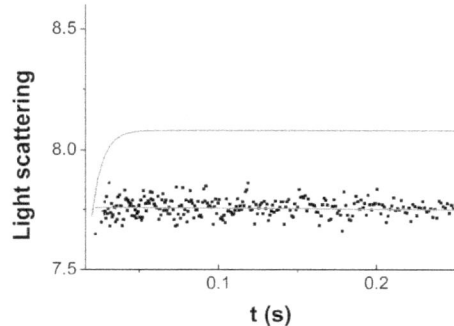

Figure 2.3: Water permeability of the vesicles determined by a stopped-flow apparatus. The increase in the light scattering signal represents a reduction in vesicle size due to water efflux from the vesicle to the solution. The black curve represent light scattering signal for the blank vesicles which are fabricated from PMOXA-PDMS-PMOXA (1000-4000-1000). The yellow curve represent light scattering signal for the aquaporin incorporated vesicles AqpZ-PMOXA-PDMS-PMOXA (1000-4000-1000). The data are exponentially fitted by equation (1). The solid lines represent the fitting curves.

Table 1: The t value shown here is used to derive k value (which is 1/t). The k value will then be incorporated into equation (1) to determine the permeability of various polymers.

t value (s)	Blank polymer vesicles (3500)	AqpZ-polymer vesicles (3500)	Blank polymer vesicles (6000)	AqpZ-Polymer vesicles (6000)
1	0.0904	0.0215	---	0.00591
2	0.10651	0.02117	---	0.00664
3	0.09362	0.01425	---	0.00947

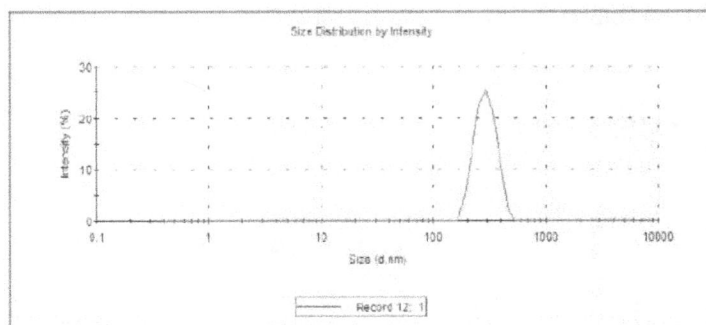

Figure 3.1: Dynamic Light Scattering graph for the aquaporin incorporated vesicles AqpZ PMOXA-PDMS-PMOXA (1000-4000-1000)

*Figure 3.1 shows the loops detected from DLS. If there is only 1 loop, it indicates that the detergent that was added together with the proteins during the experiment are successful removed by the bio beads method, and will not affect the accuracy of the results.

Table 2: Size of vesicles

Vesicles	Vesicle Size/ nm
Blank polymer vesicles (3500)	269.0
AqpZ-polymer vesicles (3500)	241.5
Blank polymer vesicles (6000)	299.6
AqpZ-Polymer vesicles (6000)	380.4

Table 3: Permeability values of vesicles

Vesicles	Permeability/ ms^{-1}
Blank polymer vesicles (3500)	8.57×10^{-5}
AqpZ-polymer vesicles (3500)	4.873×10^{-4}
Blank polymer vesicles (6000)	--
AqpZ-Polymer vesicles (6000)	1.49×10^{-3}

All the values derived (as shown on previous pages) are substituted into equation (1) to determine the permeability of vesicles (polymer with molecular weights of 3500 and 6000) with or without proteins.

After analyzing our samples using the Origin software and dynamic light scattering apparatus, it is evident that there is a visible difference between the permeability of the vesicle with a PDMS (hydrophobic block) molecular weight of 4000 and the other vesicle with a PDMS molecular weight of 2500. Results showed that PMOXA-PDMS-PMOXA with hydrophobic molecular weight of 4000 has a higher permeability value than the polymer with a hydrophobic molecular weight of 2500. The results also indicate that we

successfully incorporated of the Aquaporin Z to into the solution as it increases the permeability of the vesicles. This is especially so in the 1000-4000-1000 which reflects a higher difference in permeability value of the vesicles with aquaporin as compared to the one without.

1.4 Discussion

The dynamic light scattering results (Table 2) enabled us to find the values of vesicle diameters. With the diameters of the vesicle, we were able to calculate the surface area and volume of the vesicle, thereby substituting these values in

74

the equation to calculate its permeability. With reference to the loop shown in figure 3.1, only 1 loop is detected by the light sensor. This indicates that the detergent was completely removed from the samples from using the biobeads, giving rise to accurate results. The narrow curves also shows that the vesicle size distribution is relatively uniform, further increasing the accuracy of the permeability value (Pf). By substituting the diameter of the vesicles into equation (1), the surface area and vesicle volume of the vesicles can be obtained and these values are further substituted into the permeability equation to find Pf. The vesicle with the highest Pf value is most permeable to water molecules, thus making it more efficient in water purification.

From the scattered data on graph, there is a large variation of results generated. However, by getting the shortest time which vesicle is shrunk and light detected, we managed to determine k value (initial rate constant) and the trend of light scattering. However, the k value of blank PMOXA-PDMS- PMOXA (1000-4000-1000) vesicles is too small to be calculated, as it is inversely proportionate to the t value which is infinity. Thus, this indicates very low vesicle permeability. From Table 1.3, there is a significant increase in vesicle permeability after incorporating with aquaporin, indicating that aquaporins are successfully reconstituted into the polymer vesicles and that aquaporins indeed facilitated the transport of water molecules. In addition, with the same polymer to protein ratio, the permeability of AqpZ-PMOXA-PDMS-PMOXA (1000-4000-1000) vesicles are much higher than the permeability of AqpZ-PMOXA-PDMS-PMOXA (500-2500-500) vesicles. This indicates that the incoporation of aquaporin Z resulted in a higher permeability in a thicker vesicle.

Evident from Table 3, the permeability of vesicles incorporated with aquaporin is higher than the blank vesicles. The effect of the addition aquaporins that increases the permeability of the vesicle to water molecules is apparent in both the vesicles with a molecular weight of 3500 and 6000, but more apparent in the vesicle with the molecular weight of 6000 which permeability value increased from 0 to 1.4886 x 10-3 ms-1. This shows that the pure polymer vesicle was impermeable to water molecules initially, but the addition of aquaporins increased its permeability greatly, allowing water molecules to pass through. This is probably due to the addition of aquaporins, which create water channels that facilitate the transport of water molecules, hence increasing the permeability of the vesicles. Hence, the results of the experiment proved that the incorporation of aquaporins increases the permeability of the vesicles to water molecules. From the permeability values of the vesicles (Table 3), it is evident that the vesicle with a molecular weight of 6000 with aquaporins is more permeable to water molecules as compared to the vesicle with a molecular weight of 3500. Therefore, this shows that the thicker the vesicle, the higher the permeability value, hence resulting in a more efficient water purification process. Hence, it can be safely concluded that the experimental results proved our hypothesis right.

Although the experiment was relatively successful and produced accurate results, there are a few limitations present. One limitation is that the diameters of the vesicles are not exactly similar due to human error during the experimental procedure despite extruding them 10 times through a filter unit of 0.22 μm. Hence, the differing

diameter of the vesicles will affect the surface area and volume of the vesicles that will be substituted into the permeability equation. The permeability value of the vesicles can be affected.

Large improvements in the efficiency of water treatment membranes may result from the development of biomimetic membranes with high permeability and selectivity. The orders-of-magnitude increase in permeability observed on incorporation of AqpZ indicates that the water-channel protein is functional in the synthetic context. The magnitude of increase in permeability and the excellent solute rejection capabilities demonstrate the potential benefit of such membranes for water treatment. The protein- polymer has salt rejection and permeability ideal for desalination. Therefore, the incorporation of AQPs into compatible synthetic polymers, such as the block copolymer system investigated in this study, is an innovative approach for making membranes industrial and municipal desalting applications.

1.5 Conclusion

The experiment can be further extended by incorporating aquaporins into biomimetic membrane, which is more applicable to the water purification process as compared to vesicles due to its elongated structure. As the idea of using biomimetic membrane is relatively new in the industry, researchers have yet to discover feasible methods to improve on the poor mechanical property and stability of biomimetic membrane after the addition of AQP Thus, we would like to further our project by coming up with methods to improve the stability of biomimetic membrane together with the incorporation of AQP to improve on the current desalination process.

Acknowledgements

We would like to acknowledge and express our sincere gratitude to our mentor, Professor Tong Yen Wah for his guidance throughout the course of this project. We would also like to extend our utmost appreciation to our teacher mentor, Mr Yang Kian Hong as well as the research fellows from NUS, MsKritika Kumar and MsXie Wen Yuan, for their invaluable comments and support on our project. Last but not least, we would like to thank the lab technician, MrAng Wee Siong, for providing us with the necessary equipment for our lab work.

References

[1]   S. Loeb, S. Sourirajan, Sea water demineralization by means of an osmotic membrane, Adv. Chem. Ser. 38 (1962) 117 (et seq).

[2]   R. Truby, Chapter 4: seawater desalination by ultralow energy reverse osmosis, In: in: N. Li, N, et al., (Eds.), Advanced Membrane Technology and Applications, J.Wiley, 2008, pp. 87–100.

[3]   M.Elimelech,W.A.Phillip,The future of seawater desalination: energy, technology, and the environment, Science 333 (2011) 712–717.

[4]   Knepper MA , Wade JB , Terris J , Ecelbarger CA , Marples D , Mandon B , Chou CL , Kishore BK , Nielsen S (1996) Kidney Int 49:1712–1717.

[5]   Borgnia MJ , Kozono D , Calamita G , Maloney PC , Agre P (1999) J MolBiol 291:1169–1179.

[6]   MeinildAK ,Klaerke DA , Zeuthen T (1998) J BiolChem 273:32446–32451.

[7]   Jung JS , Preston GM , Smith BL , Guggino WB , Agre P (1994) J BiolChem 269:14648–14654.

[8]   Pontoriero G , Pozzoni P , Andrulli S , Locatelli F (2003) Nephrol Dial Transplant 18(Suppl 7):vii21–vii25.

[9]   PastanS , Bailey J (1998) N Engl J Med 338:1428–1437.

[10] Highly permeable polymeric membranes based on the incorporation of the functional water channel protein Aquaporin Z (2007) Retrieved 20 June 2013 from http://www.pnas.org/content/104/52/20719.long

[11] HongleiWang , Tai-Shung Chung , Yen Wah Tong , KandiahJeyaseelan , ArunmozhiarasiArmugam , Zaichun Chen , Minghui Hong , and Wolfgang Meier (2011) Highly Permeable and Selective Pore-Spanning Biomimetic Membrane Embedded with Aquaporin Z (2011)

Techniques of Water Purification

Schin Bek, Chloe See, Wei Xuan Tay, Vanessa Chuang

Raffles Girls' Secondary School, Singapore, shaun.desouza@rgs.edu.sg

Abstract

Our project will have a two-pronged approach to water purification, namely the removal of oil from water and the modified SODIS method of water treatment.

Firstly, we hope to remove oil from water. We aim to investigate the use and effectiveness of different recycled (reusable) materials, those that are eco friendly and biodegradable materials in removing oil pollutants from clean water. The reason why we use biodegradable materials is because they are often being thrown as trash by mankind, and they can decompose easily, thereby not polluting our environment. Our hypothesis is that these substances can help to absorb oil effectively from food and clean water.

We hope to improve on the existing process of SODIS water treatment. Through experimentation using our modified method of SODIS water treatment as well as alternative methods of water treatment, we seek to determine the most effective and environmentally-friendly techniques of water purification.

Keywords

water, purification, SODIS, biodegradable

Content

1.1 The purpose of the investigation

Water is without a doubt one of the most essential nutrients in the sustenance of human life, and water purification technology is becoming increasingly important in our world. Our project will have a two-pronged approach to water purification, namely the modified SODIS method of water treatment and the removal of oil from water.

Solar water disinfection, more commonly known as the SODIS method, is a simple, environmentally friendly, low-cost procedure to disinfect drinking water at a household level. Non-potable water is filled in a transparent PET bottle and exposed to the sun for 6 hours. SODIS uses solar energy to destroy pathogenic microorganisms causing waterborne diseases, improving the quality of drinking water. Pathogenic microorganisms are vulnerable to two effects of the sunlight: radiation in the spectrum of UV-A light (wavelength 320-400nm) and heat (increased water temperature). The SODIS method helps to prevent waterborne diseases such as diarrhoea, cholera, *E. coli* and dysentery, and thereby saves the lives of many people with no access to clean, potable water. This is urgently necessary as more than 4000 children die every day from the consequences of diarrhoea.

Despite being a fairly widespread method of disinfecting water, the SODIS method was developed in the 1990s, and no significant improvements or discoveries have been made regarding this technology since then. Many factors affect the effectiveness of SODIS water treatment, and by undertaking this project, we hope to improve on the existing process of SODIS water treatment.

For our experimental results to contribute to the existing pool of knowledge, no matter how small our contribution, we feel that we ought to take into account how the undertaking of this project could potentially help people around the world. Therefore, the rationale for this project is to make a contribution to water technology that could potentially help people in need. We hope that our discoveries through this project will be able to further the current knowledge about this topic, and that it will later have a positive impact on the spread of waterborne diseases in third-world countries.

For the second approach to water purification, we will be removing oil from water with the use of biodegradable materials. In this approach, we aim to investigate the use of different eco-friendly, biodegradable materials, by evaluating their effectiveness in removing oil pollutants from clean water.

We will be exploring the concepts of absorption and adsorption. Absorption is the process when a fluid permeates or is dissolved the absorbent. Absorption involves the whole volume of the material. Adsorption, on the other hand, is the adhesion of atoms, ions, or molecules from a gas, liquid or dissolved solid to a surface. This process creates a film of the adsorbate on the surface of the adsorbent. Adsorption is a surface-based process and it is the consequent of surface energy.

Oil is a neutral, non-polar substance. In our investigation, we will be focusing on the use of vegetable oil, which can be obtained easily. Besides, in our investigation, we will be experimenting with possible absorbents and adsorbents of oil that are biodegradable, namely egg shells and fruit peels. Egg shells are known to be lipophilic, as they absorb oil. Fruit peels, such as orange peels and banana skin, have the ability to absorb oil.

Currently, oil spills are not removed effectively and are mostly left in the sea. By undertaking this project, we hope to find effective and suitable methods of removing oil from sea so as to protect our fragile marine environment.

In conclusion, we hope that our project will be able to contribute to existing water purification technology, and that this will in turn allow for increased water security across the globe.

1.2 Method of the investigation

With regards to the alternative water treatment methods when sunlight is scarce, we will be conducting several experiments to investigate the effectiveness of three alternative methods of water treatment. A high (>7) or low pH (<7), ultrasonic irradiation and the addition of certain chemical compounds may inhibit or minimize the growth of bacteria in general. For the experiment on pH, the hypothesis is that the greater the deviation of pH from a neutral pH of 7, the more effective the water treatment. For the experiment on ultrasonic irradiation, we predict that it will be effective in treating the water. Lastly, for the experiment on the addition of chemical compounds, the hypothesis is that certain chemical compounds added will have antimicrobial properties.

An example of our experimental procedure is as follows – this is the methodology for an alternative method of water treatment, in which the pH is acidic.

Step	Task
1	Obtain 6 150ml beakers. Fill them with the previously mentioned water-probiotic solution of equal volumes. (Negative control, positive control, original, 3 samples) Take pH readings for each sample before and after the experiment.
2	Immediately culture 100 microlitres of solution from the 'original' sample on an agar plate.
3	Do not add anything to the negative control, and let it stand for 15min.
4	Add bleach to the positive control, shake it and let it stand for 15min.
5	Add an equal volume of solution (e.g. freshly squeezed lemon juice) into the remaining 3 samples. Let them stand for 15min.
6	Once 15min has passed, culture 100 microlitres of solution from each sample on an agar plate.
7	After 48 hours, perform a live bacterial colony count and record all data down.

Our data analysis for the experiments is as follows - the extent to which each factor affects the effectiveness of the SODIS water treatment would be determined by the number of live bacteria colonies at the end of the experiment. This would be obtained by using the method of bacteria culture streaking, hence allowing the reproduction of bacteria in a controlled environment. The greater the number of bacteria colonies, the less effective the method of SODIS water treatment.

Methodology for finding out the effectiveness of absorbents/adsorbents in removing oil spills by determining the percentage change in mass of the solid* used.

*We shall investigate the effectiveness of different recycled/reusable biodegradable materials (example- fruit peels etc.)

For absorbents/adsorbents that sink in water:

Step	Task
1	Measure the mass of the empty beaker, M_1, which is going to be used in the experiment. Record the reading in a table.
2	Decide on the volume of the oil and water that is going to be used during the experiment. Record the initial volumes of the oil and water. Ensure that the volume of both oil and water are the same for all set-ups.
3	Add the same mass of solid into the set-ups and leave the substance in each setup for a pre-determined amount of time.
4	Decant the mixture of oil and water into another beaker, leaving the solid behind, with some oil sticking onto its surface.
5	Measure the mass of the beaker, which now contains the solid. Record the mass, M_2.
6	Subtract M_1 from M_2 to find the mass of the solid after the experiment.
7	Repeat steps 1 to 6 for all the other setups. Repeat the entire experiment for all setups again for at least three times to ensure accuracy, reliability and validity of results.

1.3 Results of the experiment

We are still in the process of experimentation, and are unfortunately unable to provide conclusive results at present. Once we have obtained the necessary results, we will update our report accordingly.

2. Conclusion

We are unable to make a suitable conclusion right now, but this will be updated as soon as possible. Once again, the ultimate goal of our project is that it will allow us to contribute to existing water purification technology, and that this will in turn allow for increased water security across the globe.

Acknowledgements

We woud like to extend our deepest gratitude to Mr Teoh Chin Chye Alex and Mr Shaun Gerad De Souza, our mentors for this project.

References

[1] "Projects Involving Potentially Hazardous Biological Agents." *Projects Involving Potentially Hazardous Biological Agents*. N.p., n.d. Web. 10 Apr. 2014.

[2] "Solar Water Disinfection." *Wikipedia*. Wikimedia Foundation, 04 Aug. 2014. Web. 10 Apr. 2014.

[3] "How Does It Work?" *SODIS:*. N.p., n.d. Web. 10 Apr. 2014.

[3] "Is My Probiotic Supplement Active? Easy At-Home Test!" *Is My Probiotic Supplement Active? Easy At-Home Test!* N.p., n.d. Web. 10 Apr. 2014.

Green roofs: A water quality study

Ng Wei Chin, Phuah Hui Qi Phyllis, Abigayle Ng, Grace Lim

Raffles Institution Lane, Singapore, ngwc1996@gmail.com

Abstract

This study elucidated the green roofs' ability to delay rainwater flow, water retention capabilities, and the effects of green roofs on runoff water quality. A number of green roof assemblies were constructed using two substrates, garden soil and rooftop soil, and zeolite layers. Three simulated rainwater events were considered in this study. Results showed that green roof assemblies had water retention capabilities of up to 40%, and an average saturation time of 243 s. Rooftop soil is a better media for green roofs, as these assemblies generally had a longer saturation time, greater water retention capabilities and lower ion concentrations in runoff. Zeolite does not appear to be a useful filter as it exacerbates ion deposition into runoff. Overall, urban planners in Singapore are recommended to further tap on green roofs' potential to decrease water flow in drainage systems during peak rainfall, hence mitigating flooding problems in urban areas.

Keywords

Green roofs, water quality, zeolite

1   Introduction

Green roofs are gaining popularity in Singapore for reasons of practicality and aesthetic quality, even though they require high expenditure and incur maintenance costs. This is exemplified by the recent inclusion of green roofs into buildings like the Marina Barrage, Fusionopolis and Treelodge@Punggol.

Green roofs have been utilised in storm-water management [1] to delay initial runoff and reduce total runoff. For example, the extensive greening of just 10% of buildings in Brussels resulted in a runoff reduction of 54% for each building and 2.7% for the entire region [2]. Furthermore, green roofs mitigate the urban heat island effect [3]. Green roofs increase the amount of water that participates in evapotranspiration, photosynthesis and respiration, which ultimately reduce the external surface temperature of buildings [4] and lower their energy requirements [5], hence saving costs. Green roofs also improve air and water quality [6].

Despite extensive research into green roofs, the water quality of green roof runoffs has been largely overlooked [7]. Studies have warned of the environmental harm polluted runoff can cause [8], and reported vegetated roofs as a source of contaminants [9]. The great variability of roof runoff quality has also been separately noted [10],

suggesting a need to further investigate the water quality of green roof runoffs.

This study aims to assess the quality of green roof runoffs in relation to freshwater standards, and to determine the effect of different media on runoff quality. Since there are no runoff quality standards in Singapore, the relevant results of this study were juxtaposed to the requirements for freshwaters by the US Environmental Protection Agency [7, 11, 12].

2   Materials And Methods

2.1   Green roof assemblies

A typical green roof consists of four layers: vegetation, soil substrate, a geotextile filter preventing substrate loss and drainage material [13]. For this study, 16 clear acrylic assemblies of 40cm by 40cm by 25cm were used to simulate a small section of a green roof (refer to Figure 1). The assemblies were constructed with a spout at the base where rainwater runoff could flow out from.

Each assembly was filled with 3 to 5 layers. The topmost vegetation layer featured *Hemigraphis alternata*, which was selected as it is a low-maintenance plant [14], that being native to Java, Indonesia, can withstand tropical weather. This makes it suitable for use in Singapore. Also, it does not grow beyond 10cm, accommodating the height restriction of our rainwater simulation equipment. The second layer, the substrate, consisted of either local garden soil or lightweight rooftop soil (both commercially available). The use of 2 different substrates allowed for comparison on which substrate is more suitable to be used commercially in green roofs, in terms of water quality and quantity control. The next layer (where present) consisted of zeolite (3cm), an established water-moderating material used in agriculture and planting [15, 16, 17]. Zeolite is a hydrated crystalline tectoaluminosilicate constructed from TO_4, where T represents a tetrahedral interstitial atom [18]. The fourth layer was a geotextile filter cloth that is commonly used in green roofs to hold small particles in place, preventing them from being washed out of the system [7]. This allowed investigation on zeolite's ability to delay runoff, regulate runoff volumes and alter the ionic composition of runoff. The bottom drainage layer consisted of leca. The controls were identical to the other assemblies, but featured no plants. Table 1 shows the different types of assemblies constructed.

Table 1: The green roof assemblies. Each type of assembly had a setup that did not have plants and was tested with high intensity of rainfall.

| Type of assembly | Layers | | | | | No. of assemblies | Intensities of rainfall tested |
	Plants (up to 10cm)	Type of substrate (10cm)	Zeolite (3cm)	Geotextile filter	Leca (5cm)		
GN	Present	Garden	Absent	Always present	Always present	4	High, Medium, Low
GZ	Present	Garden	Present			4	High, Medium, Low
RN	Present	Rooftop	Absent			4	High, Medium, Low
RZ	Present	Rooftop	Present			4	High, Medium, Low

2.2 Experimentation and data collection

Hemigraphis alternata plants were procured from a local plant supplier and planted in the various assemblies. The assemblies were placed in a shaded area with sunlight, and watered every weekday for 17 weeks, with 400ml on Mondays and Fridays, and 300ml on Tuesdays, Wednesdays and Thursdays.

A total of 3 rainfall events were simulated on 10/12/13 (Day 1), 11/12/13 (Day 2) and 12/12/13 (Day 3) respectively. The rainwater simulator was situated at Van Kleef Centre @ Sungei Ulu Pandan. A schematic of the rainwater simulator and how it works is reflected in Figure 1 below.

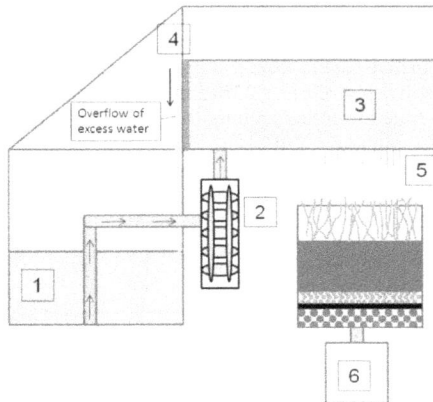

Figure 1: A schematic of the rainwater simulator and an assembly. The storage tank on the left (1) was filled with rainwater taken from the Van Kleef Centre collection systems, and then was transferred via a filter pump (2) into an overhead container (3). A

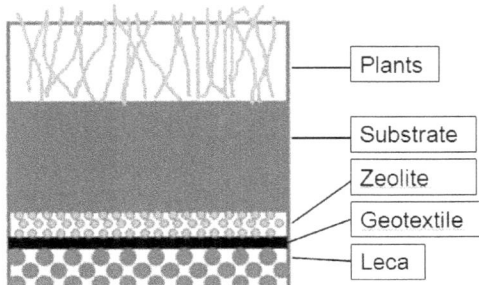

mechanism (4) at the side of the overhead container allowed for the water level in the container to be adjusted, which in turn affected the intensity and volume of rainfall onto the assembly

below. 169 micro-syringes (5), arranged in a 13x13 formation, allowed the water to form rain-like drops. Runoff is collected in a pail (6) via a tube attached to the spout.

Different intensities of rainfall were simulated for each set of four assemblies (GN, GZ, RN, RZ) and they were assigned 5cm, 10cm and 15cm water levels respectively, corresponding to 20.6mm (high), 15.6mm (medium) and 10.3mm (low) of rainfall. The three intensities of rainfall all fall under the 'heavy rainfall' category in Singapore [19]. The controls all used high intensity rainfall.

The time taken for the first drop of water to enter the pail was taken as the saturation time, t_s. Following this, water samples were taken after the 1st 5 min and after the 2nd 5 min. Turbidity readings were obtained using the Turbidity Orion AQ 4500. Conductivity, TDS, salinity and pH readings were obtained using the YSI-556 MPS. Concentrations of various anions and cations were analysed using Dionex Ion Chromatography ICS-1100, equipped with AS-DV automated sampler was used. IonPac_AS23 column in conjunction with IonPac_AG23 guard column and ASRS300 (4 mm) suppressor was used for anion measurements; IonPac_CS12A column in conjunction with IonPac_CG12A guard column and CSRS300 (4 mm) suppressor was used for cation measurements. Control water samples were taken from the reservoir of rainwater and tested using the same equipment.

3 Results and Discussion

Only results from the assemblies which had high intensity rainfall simulation are presented as representative of the worst case scenario. Please see the appendix for the other data collected.

3.1 Quality of runoff

3.1.1 pH

From Figure 2A, GN and GZ increased the pH of the runoff from 5.7-6.4 to 7.4-8.0. RN increased the pH marginally by around 0.5, while RZ did not produce any noticeable changes. In addition, the pH levels had an upward trend for garden soil assemblies while the pH remained relatively constant for rooftop soil assemblies. Garden soil assemblies may neutralise acid depositions from the rain, causing an increase in pH of runoff [9]. This is an important property, as it improves the quality of runoff that flows into water management systems. A comparison between the control and heavy rainfall results suggests that *Hemigraphis alternata* does not alter the pH of runoff.

3.1.2 Conductivity

From Figure 2B, all four assemblies increased the conductivity of runoff from 0.017 mS/cm to a maximum of 3.55 mS/cm. Garden soil assemblies had significantly higher conductivity readings than rooftop soil assemblies, indicating that they have higher ion concentrations. In addition, assemblies with zeolite had elevated conductivity readings in comparison to those without zeolite. Zeolite appears to contribute to the increased conductivity of runoff, which can be attributed to its natural composition of ions such as Na^+ and K^+ [18]. It is also evident from Figure 2B that runoff collected decreased in ion concentration with each subsequent rainfall event. This phenomenon of first flush, where elevated conductivity readings are recorded during the first rainfall and slowly decrease with subsequent events, was observed [7]. Lastly, the conductivity readings for assemblies with plants were consistently lower than those without plants by 0.71 mS/cm on average. This suggests that *Hemigraphis alternata* reduces the overall concentration of ions in runoff, possibly by absorbing nutrient ions for plant growth, which is ideal for improving runoff water quality.

3.1.3 Anions

On Day 1, the average increase in concentration of anions (relative to control readings) was 6.5 mg/L for NO_3^-, 642.6 mg/L for SO_4^{2-} and 380.6 mg/L for Cl^-. From Figure 2C and 2D, the four assemblies were significant sources of SO_4^{2-} and Cl^-. The release of SO_4^{2-} displays the first flush effect, where SO_4^{2-} content in the runoff from the 1st 5 min of Day 1 was consistently the highest, with the exception of GN. The decline in SO_4^{2-} content in subsequent runoffs, however, is not very significant. In addition, the runoff from garden soil assemblies has a greater concentration of SO_4^{2-} and Cl^- as compared to rooftop soil assemblies, averaging 707 mg/L and 235 mg/L more respectively. This can possibly be attributed to the composition of garden soil. Zeolite increased the amount of SO_4^{2-} and Cl^- in the runoff by

77% and 76% in garden soil assemblies and 65% and 12% in rooftop soil assemblies respectively. Hence, zeolite negatively affects water quality by serving as a possible source of Cl^- and SO_4^{2-}. GN, GZ and RZ were also sources of NO_3^- (see Figure 2E). Studies have reported the use of zeolite to improve the amount of available nitrogen [15], however, in this study zeolite did not affect nitrate runoff amounts. This suggests that zeolite might aid in the retention of nitrate ions. For the first two rainfall events, the runoff concentration of PO_4^{3-} increased on average by 4.31 mg/L for RN and RZ. Rooftop soil appears to be a source of PO_4^{3-}. However, the runoff collected for the final rainfall event showed negligible amounts of PO_4^{3-}. Lastly, comparing the results of assemblies with plants and assemblies without plants, *Hemigraphis alternata* decreases the concentration of Cl^- by 79.8 mg/L on average, possibly because it takes in Cl^- as nutrients for plant growth. There was no difference in the concentration of ions such as SO_4^{2-} and NO_3^- in assemblies with and without plants.

3.1.4 Cations

On Day 1, the average increase in concentration of cations (relative to control readings) was: 195.6 mg/L for Na^+, 102.7 mg/L for K^+, 248.8 mg/L for Ca^{2+} and 35.2mg/L for Mg^{2+}. All four assemblies were sources of Na^+, K^+, Ca^{2+} and Mg^{2+}. The release of Mg^{2+} and K^+ generally display the first flush effect as mentioned above. However, the decline in ion concentration in subsequent runoffs is not very significant. Runoff from garden soil has a greater concentration of cations than rooftop soil, especially for Ca^{2+}, where the difference in average readings for garden soil and rooftop soil assemblies was 306 mg/L. In addition, zeolite increased the concentrations of Na^+ and Mg^{2+} in garden soil assemblies by 27% and 152% respectively. This could be due to zeolite's ability to exchange these ions with contact solution [20]. GN had consistently lower concentrations of NH_4^+ than control readings, suggesting that it served as a sink for NH_4^+. This was also observed for some readings of RN and RZ.

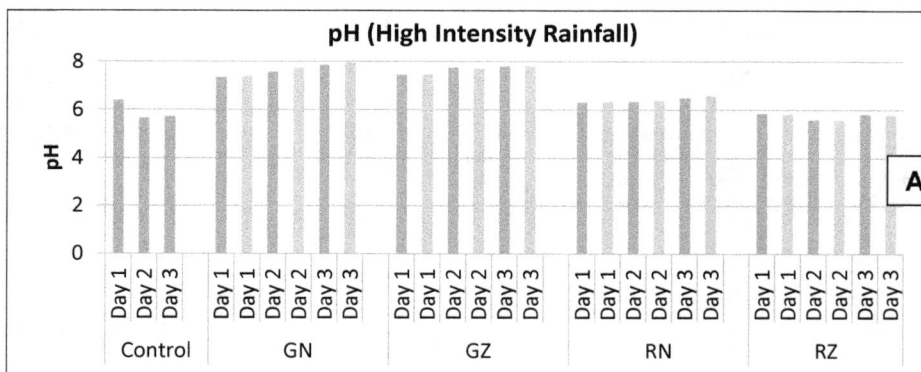

Legend: ■ - 1st 5 min ■ - 2nd 5 min

Figure 2A: pH of runoffs during high intensity rainfall simulation.

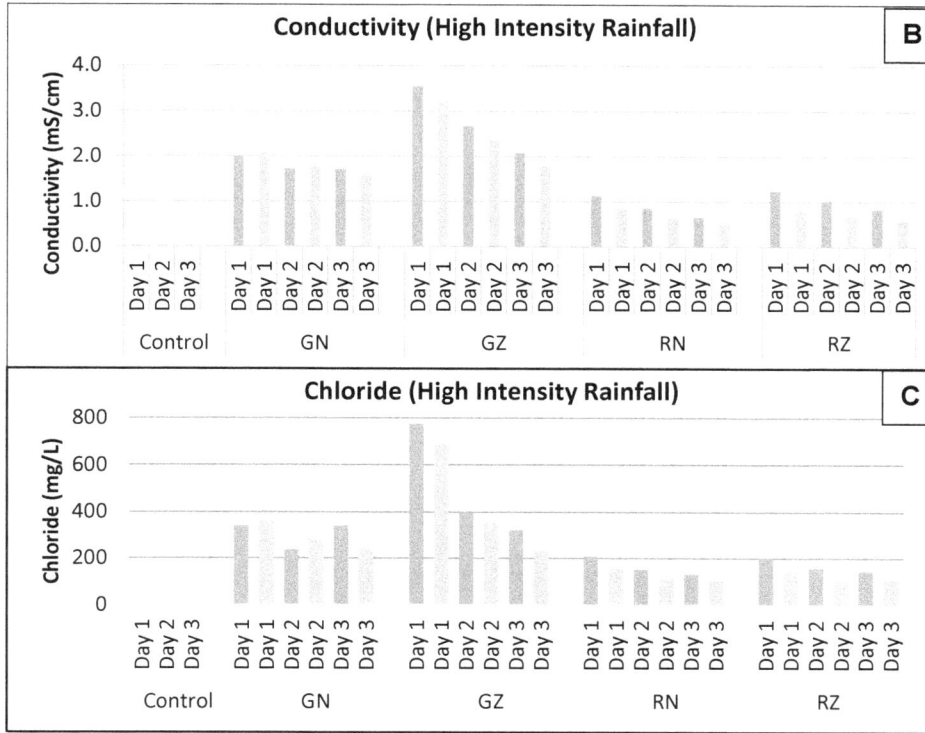

Figure 2B: Conductivity of runoffs during high intensity rainfall simulation.

Figure 2C: Amount of Cl⁻ in runoffs during high intensity rainfall simulation.

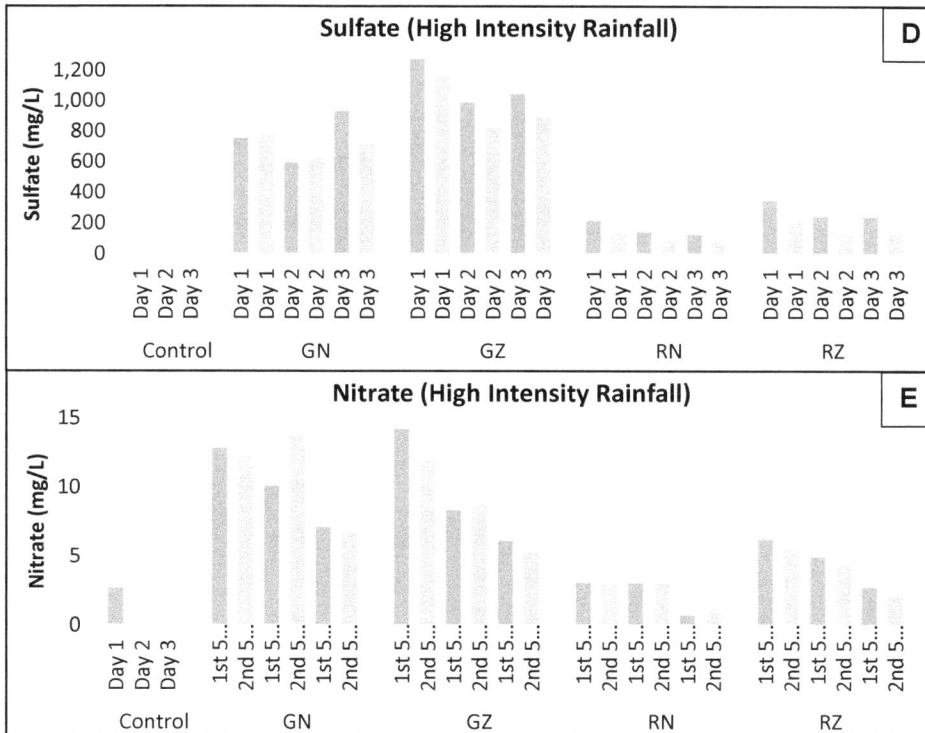

Figure 2D: Amount of SO_4^{2-} in runoffs during high intensity rainfall simulation.

Figure 2E: Amount of NO_3^- in runoffs during high intensity rainfall simulation.

3.2 Other aspects of the runoff

3.2.1 *Saturation time, t_s*

RZ consistently had the longest t_s, followed by RN, GZ and finally GN. In general, rooftop soil assemblies have longer t_s (averaging 275 s) as compared to garden soil assemblies (212 s). This could possibly be attributed to the closely packed nature of garden soil which may increase moisture retention. The presence of a zeolite layer also delays t_s. The t_s of RZ was 140 s longer than RN while the t_s of GZ was 33 s longer than GN. This suggests the role that zeolite may play in storm water management by delaying the initial runoff time. Also, consecutive rainfall events appear to affect t_s, as t_s of the initial rainfall event was consistently higher than that of the subsequent events for each assembly.

3.2.2 *Water retention capabilities*

The assemblies reduced the volume of runoff by approximately 10%, indicating their ability to reduce peak water flow and this is independent of the type of substrate. In general, the volume of runoff collected is consistently higher during the second 5 min than the first. Zeolite did not significantly reduce the volume of runoff, even though is known for its dehydration properties [15]. Zeolite has a porous polyhedral structure, which provides void volumes for water absorption. This could perhaps be attributed to the 17 weeks of prior watering which greatly reduced the water-moderating capabilities of zeolite.

Across the four assemblies, overall water retention capabilities ranged from 29%-40% in the (ascending) order: GN, RN, GZ, RZ. Overall water retention capability was calculated using Eq.(1).

$$Overall\ water\ retention\ capability = \frac{(3300 + 5.5\,S) - O}{3300 + 5.5\,Ts} \times 100\% \qquad (1)$$

where O is the observed volume of runoff in 10 min in cm^3;

   t_s is the saturation time in s;

   3300 is the amount of rainfall in 10 min, in cm^3; and

   5.5 is the rate of rainfall in cm^3/s.

A modified formula was used to calculate the respective water retention capabilities for each 5 min interval. It was observed that the water retention capabilities for the 2nd 5 min were on average 7% lower than the 1st 5 min. This indicates that as the substrate gets saturated with water, its water retention capabilities are reduced [7]. Given that the experiment simulated rainfall for only 10 min, this is reflective of a possible weakness of green roofs in alleviating storm-water drainage problems during intense rainfall over longer periods of time, as the water retention capabilities of the green roofs steadily decreases. Lastly, contrary to most research [21], the presence of vegetation appeared to reduce the water retention capabilities of the substrate instead of increasing it.

3.2.3 Physical appearance

While turbidity and colour does not indicate actual chemical concentration, it affects aesthetic appeal. GN had the highest average turbidity readings of 45.6 ppm, followed by RZ, GZ and lastly RN with a reading of 12.7 ppm. Runoff collected during the 1st 5 min was generally more turbid than 2nd 5 min. This is indicative of the first flush effect. The presence of a zeolite increases turbidity in rooftop soil assemblies but not garden soil assemblies. In addition, assemblies with plants are consistently less turbid than assemblies with plants than without plants. This suggests that *Hemigraphis alternata* helps to reduce the amount of suspended solids in runoff. Similarly for colour, runoff collected from garden soil assemblies was generally clear in comparison to the clear brownish runoff from rooftop soil assemblies. Colour gives the wrong impression of the dirtiness of runoff, as while runoff from rooftop soil assemblies appear more dirty, ion concentration readings above show that runoff from garden soil assemblies is generally more polluted with ions.

3.3 Overall evaluation

Many of the parameters studied exceeded the limits for the USEPA freshwater standards. The amounts of SO_4^{2-} exceeded the standards across all the assemblies, while those for NO_3^- and Cl^- only exceeded the limits in the garden soil assemblies. In contrast, the pH levels and PO_4^{3-} standards only exceeded the limits for the rooftop soil assemblies, and the F^- amounts were acceptable for all the assemblies. In general, the green roof assemblies broadly act as contaminants. Considering the results in Table 2, rooftop soil is the preferred substrate if the priority is reducing contamination.

Table 2: Comparison of runoff with USEPA recommended freshwater standards.

Parameter	Recommended Standards	GN	Pass limit	GZ	Pass limit	RN	Pass limit	RZ	Pass limit
pH	6.5-9	7.34-7.94	No	7.44-7.82	No	6.30-6.56	Yes	5.58-5.85	Yes
NO_3^-	10	6.67-13.7	Yes	5.22-14.2	Yes	0.70-3.03	No	2.06-6.23	No
PO_4^{3-}	0.05	N.A.	N.A.	N.A.	N.A.	0.29-4.54	Yes	0.08-4.37	Yes
SO_4^{2-}	250	591-932	Yes	821-1275	Yes	91.4-215	No	131-346	Yes
Cl^-	230	236-363	Yes	230-774	Yes	103-209	No	108-203	No
F^-	4.0	0.49-0.89	No	0.47-1.04	No	0.44-1.0	No	0.37-0.95	No

4   Conclusion

It is recommended that green roofs utilise rooftop soil as its runoff has a lower ion concentration. It is also able to delay and reduce peak flow. Zeolite appears to exacerbate pollutant deposition into runoff, with elevated levels of conductivity, Cl^- and SO_4^{2-} observed in the assemblies containing zeolite. It delays saturation time marginally, and does not noticeably reduce volume of runoff. Hence, zeolite is dismissed as a useful filter in constructing green roofs; however, there is no conclusive proof of its detriments. Further research should be undertaken regarding zeolite and other filters to ascertain their suitability for widespread application in green roof construction. It is also necessary to find plants with better remediative capabilities than *Hemigraphis alternata,* such that the plants are better able to absorb water and nutrients, such that runoff quality meets USEPA's recommended freshwater standards, which would enhance the effectiveness of green roofs.

Positive aspects of green roofs found are their ability to delay runoff time and reduce the volume of runoff. When implemented in densely urbanised settings, green roofs may play a large role in stormwater management. Also, it serves as an aesthetically appealing green lung that beautifies its concrete surroundings. Urban planners and policy makers should tap on the potential of green roofs to tackle recent stormwater management issues that have been recurring in Singapore.

ACKNOWLEDGEMENTS

The authors gratefully acknowledge the support of Van Kleef Centre @ Ulu Pandan towards the research activities carried out, and Dr Joshi Umid Man for the liaison and guidance. The authors also thank the Raffles Science Institute and Raffles Institution (Junior College) for providing research facilities for use, and Dr Abigayle Ng and Dr Grace Lim for the guidance throughout this research.

REFERENCES

[1] Oberndorfer E., Lundholm, J., Bass, B., Coffman R. R., Doshi H., Dunnett N., Gaffin S., Köhler M., Liu, K. K. Y., and Rowe, B. (2007). Green Roofs as Urban Ecosystems: Ecological Structures, Functions, and Services. BioScience 57(10).

[2] Mentens, J., Raes, D., and Hermy, M. (2006). Green roofs as a tool for solving the rainwater runoff problem in the urbanized 21st century? Landscape and Urban Planning 77(3), 217-226.

[3] Bass, B., Stull, R., Krayenjoff, S., and Martilli, A. (n.d.). Mitigating the Urban Heat Island with Green Roof Infrastructure.

[4] Niachou, A., Papakonstantinou, K., Santamouris M., Tsangrassoulis A., and Mihalakakou, G. (2001). Analysis of the green roof thermal properties and investigation of its energy performance. Energy and Buildings 33(7), 719-729.

[5] Onmura, S., Matsumoto, M., and Hokoi, S. (2001). Study on evaporative cooling effects of roof lawn gardens. Energy and Buildings 33(7), 653-666.

[6] Wong, N. H., Tay, S. F., Wong, R., Ong, C. L., and Sia, A. (2003). Life cycle cost analysis of rooftop gardens in Singapore. Building and Environment 38(3), 499-509.

[7] Vijayaraghavan, K., Joshi, U. M., Balasubramanian, R. (2012). A field study to evaluate runoff quality from green roofs. Water Research 46 (2012), 1337-1345.

[8] Getter, K. L., and Rowe, D. B. (2006). The Role of Extensive Green Roofs in Sustainable Development. HortScience 41(5), 1276-1285.

[9] Berndtsson, J. C., Emilsson, T., Bengtsson, L. (2006). The influence of extensive vegetated roofs on runoff water quality. Science of the Total Environment 355 (2006), 48-63.

[10] Vialle, C., Sablayrolles, C., Lovera, M., Jacob, S., Huau, M.-C., Montrejaud-Vignoles, M. (2011). Monitoring of water quality from roof runoff: Interpretation using multivariate analysis. Water Research 45 (2011), 3765-3775.

[11] U.S. Environmental Protection Agency (USEPA), 1986. Quality Criteria for Water 1986. Office of Water, Regulation and Standard, Washington, DC. 20460 (2006).

[12] U.S. Environmental Protection Agency (USEPA), 2009. National Recommended Water Quality Criteria. Office of Water, Office of Science and Technology, Washington DC. 4304T (2009).

[13] Berndtsson, J. C. (2010). Green roof performance towards management of runoff water quantity and quality: A review. Ecological Engineering 36 (2010), 351-360.

[14] Missouri Botanical Garden. (n.d.) Hemigraphis alternata. Retrieved from
http://www.missouribotanicalgarden.org/PlantFinder/PlantFinderDetails.aspx?kempercode=a514

[15] Mumpton, F. A. (1985). Using Zeolites in Agriculture. Department of the Earth Sciences, State University College, Brockport, NY 14420.

[16] Mumpton, F. A. (1999). Uses of natural zeolites in agriculture and industry. Proceedings of the National Academy of Sciences of the United States of America 96(7), 3463-3470.

[17] Rydenheim, L. (2007). Effects of zeolites on the growth of cucumber and tomato seedlings. Horticultural Science programme 2007-04, 10 p (15 ECTS).

[18] Davis, M. E., and Lobo, R. F. (1992). Zeolite and Molecular sieve synthesis. Chemistry of Materials 4(4), 756–768.

[19] National Environmental Agency. (2008). Weather Notes. Retrieved from

http://www.weather.gov.sg/online/loadNotesProcess.do

[20] Curkovic, L., Cerjan-Stefanovic, S. and Filipan T. (1997). Metal ion exchange by natural and modified zeolites. Water Research 31 (6). 1379-1382.

[21] Van Dijk, P. M., Kwaad, F. J. P. M., Klapwijk, M. (1998). Retention of Water and Sediment by Grass Strips. Hydrological Processes 10(8), 1069-1080.

Impending Flood-effects in 'The Baardwijkse Overlaat'

Ellis van Dijk, Lisa Kolmans

d'Oultremontcollege Drunen, The Netherlands, p.vankempen@doultremontcollge.nl

Abstract

In their struggle against water, the Dutch government have assigned areas to absorb the excess water during impending floods. One of these areas is 'De Baardwijkse Overlaat' which is characterized by a high biodiversity.

This investigation is taking place around two separate pools of water with a difference in biodiversity and will last for at least one year. By frequently examining types and levels of nutrients in surface waters and soil, the students will try to clarify the impact which nutrients (nitrate, nitrite, sulfate, phosphate, iron, carbon dioxide, oxygen and pH-value) have on biodiversity.

From this progressive research, seasonal abiotic factors can be identified by monitoring the amount of nutrients in the environment. Interpreting the results can confirm a possible correlation between fluctuations in nutrient levels in the environment related to the composition of these within the producers (plants).

Students will get acquainted with techniques like chromatography and photo spectrometry.

Keywords

Biodiversity, nutrient levels, abiotic factors, iron, phosphate

1   Introduction

In January 2014 we started our investigation. At this time of year, it is winter in The Netherlands and nature is asleep. We started out by finding out how to determine the amount of iron and phosphate in water samples. This forms the main part of this report.

At this moment nature is awakening and the investigation of the biodiversity is about to begin. At the 'Water is Life International Youth Conference' we will tell you more about this part of our investigation.

2   Content

2.1   The purpose of the investigation

Iron and phosphate are two important nutrients involved in the growth of plants. The first part of our investigation is to determine the amounts of these nutrients in water samples of two different pools and their surrounding soil.

Iron is essential for all plants. It is indispensable for the formation of chlorophyll. Chloroplasts consist for about 80% of iron. It is therefore essential for photosynthesis. Iron is a component of proteins and enzymes. The plants take up iron in the form of Fe^{2+} ions and therefore the Fe^{3+} ions in the root surface should be reduced, so they can be incorporated. In grasses the Fe^{3+} ions are the most important ones.

Iron is an element that is common in surface- or groundwater. Iron is also non-toxic (for fish), except when the levels are very high. The iron ions are originate from iron salts. These are usually highly soluble. Iron salts can decrease the pH of weakly buffered water. Subjacent water will accelerated the precipitation of $Fe(OH)_3$. This turns the water cloudy and it has a typical rust brown colour. A low pH, or a lot of air will increase the content of soluble iron. A high pH, a low ground temperature or poor root growth reduce the absorption of iron.

Phosphor is an important plant nutrient, making up about 0,2% of a plant's dry weight. It's a component of key molecules as nucleic acid, phospholipids and ATP. Plants cannot grow without a reliable supply of this nutrient.

In biological systems phosphorus is found as free phosphate ions, PO_4^{3-}. Phosphate compounds, together with inorganic nitrates, are the main ingredient of fertilizers. The amount of phosphates and nitrates in water are very important, too much of both ions in surface water (caused by excessive fertilization of the environment) will lead to a lack of oxygen (eutrophication).

2.2   Method of the investigation

2.2.1 Determining environmental conditions

Environmental conditions relevant to our investigation are temperatures en pH values of the water samples.

Also the average temperature and the amount of rainfall in the area will be monitored.

2.2.2 Determining the amount of iron

Spectrophotometric techniques have been used to measure the concentration of Fe^{3+} in six calibration samples and several test samples.

The standard solution (10 mg Fe^{3+} L^{-1}) was made by dissolving 72 mg of iron(III)nitrate-nona-hydrate and 20 mL of 37% HCl in distilled water up to a volume of 1.0 L.

The six calibration samples were made by mixing the standard solution (A) with 2M potassiumthiocyanate (B) and distilled water (C).

Nr	A (mL)	B (mL)	C (mL)
1	0.0	1.0	9.0
2	1.0	1.0	8.0
3	2.0	1.0	7.0
4	3.0	1.0	6.0
5	4.0	1.0	5.0
6	5.0	1.0	4.0

Table 1

The test samples are made by mixing 5 mL of water (from the different pools) with 1.0 mL 2M hydrochloric acid, 1.0 mL 2M potassiumthiocyanate solution and 3.0 mL of distilled water.

The extinction of the calibration samples and test samples are measured by a photo spectrometer at 480 nm.

2.2.3 Determining the amount of phosphate

Spectrophotometric techniques have also been used to measure the concentration of PO_4^{3-} in six calibration samples and several test samples.

The standard solution (30 mg PO_4^{3-} L^{-1}) was made by dissolving 107 mg of potassium phosphate-hepta-hydrate in 1.00 L of distilled water.

Reagent I: add 400 mg ammoniuim-hepta-molybdate to 100 mL 0,5 M sulphuric acid.

Reagent II: add 2 mg 1-amino-2-naftol-4-sulfonicacid and 200 mg sodiumsulphite to 100 mL of distilled water.

The six calibration samples were made by mixing the standard solution (A) with reagent I (B), reagent II (C) and distilled water (D).

Nr	A (mL)	B (mL)	C (mL)	D (mL)
1	0.0	3.0	2.0	5.0
2	1.0	3.0	2.0	4.0
3	2.0	3.0	2.0	3.0
4	3.0	3.0	2.0	2.0
5	4.0	3.0	2.0	1.0
6	5.0	3.0	2.0	0.0

Table 2

The test samples are made by mixing 5 mL water (from the different pools) with 3.0 mL reagent I and 2.0 mL reagent II.

Fifteen minutes after mixing, the extinction of the calibration samples and test samples are measured by a photo spectrometer at 352 nm.

2.3   Results of the experiment

2.3.1 Environmental conditions

The graphs below show the environmental conditions during our investigation.

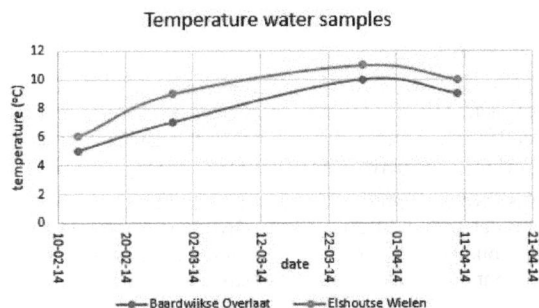

Figure 1

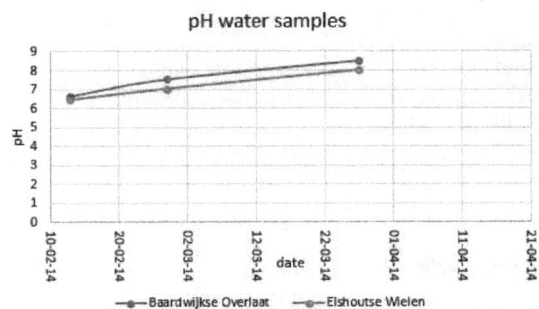

Figure 2

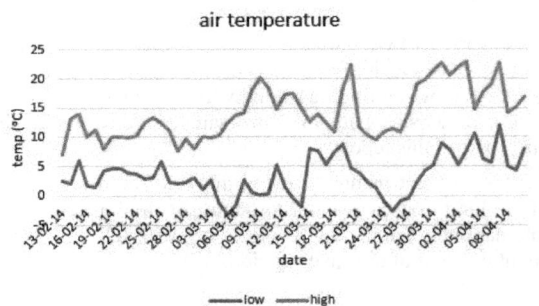

Figure 3

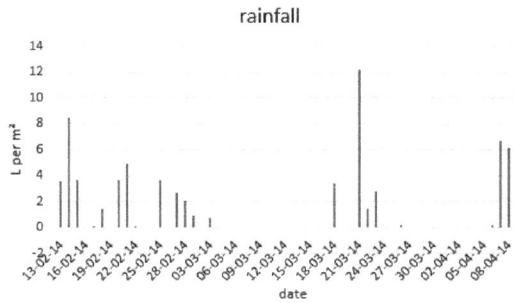

Figure 4

2.3.2 Investigating the amount of Iron

To determine the amount of iron you first need a calibration curve.

Calibration sample	Fe^{3+} (mg L^{-1})	Extinction
1	0.0	0.000
2	1.0	0.149
3	2.0	0.315
4	3.0	0.490
5	4.0	0.683
6	5.0	0.791

Table 3

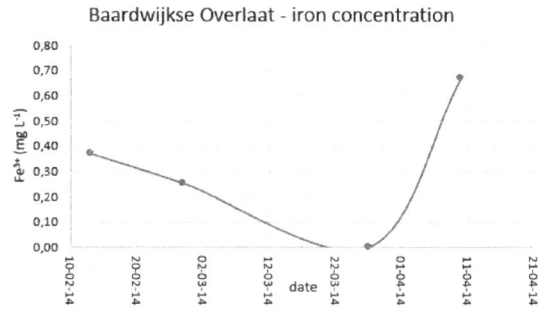

Figure 5

Test samples Baardwijkse Overlaat

Date	Extinction	Fe^{3+} (mg L^{-1})
13-02-14	0.060	0.37
27-02-14	0.041	0.25
27-03-14	0.000	0.00
10-04-14	1.109	0.67

Table 4

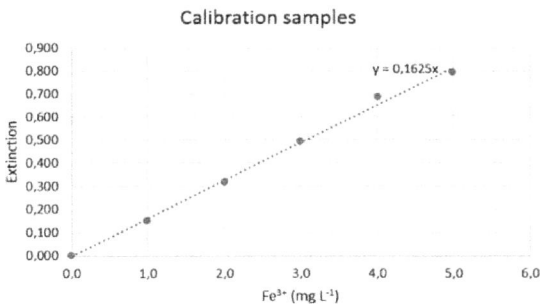

Figure 6

Test samples Elshoutse Wielen

Date	Extinction	Fe^{3+} (mg L^{-1})
13-02-14	0.026	0.16
27-02-14	0.061	0.38
27-03-14	0.125	0.77
10-04-14	0.020	0.12

Table 5

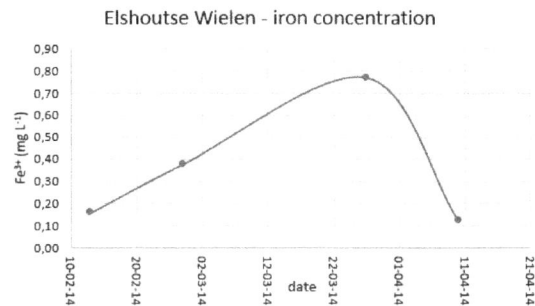

Figure 7

2.3.3 Investigating the amount of phosphate

To determine the amount of phosphate you first need a calibration curve.

Calibration sample	PO_4^{3-} (mg L^{-1})	Extinction
1	0.0	0.000
2	3.0	0.361
3	6.0	0.718
4	9.0	0.920
5	12.0	1.181
6	15.0	1.501

Table 6

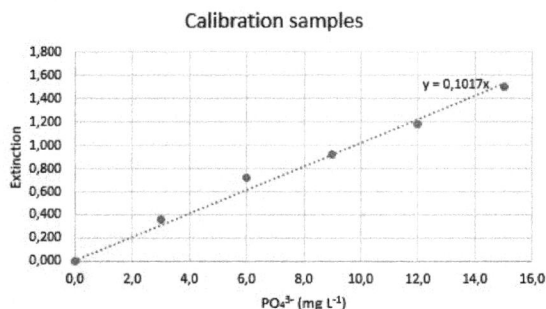

Figure 8

Test samples Baardwijkse Overlaat

Date	Extinction	PO_4^{3-} (mg L^{-1})
13-02-14	0.140	1.38
27-02-14	0.650	6.39
27-03-14	0.000	0.00
10-04-14		

Table 7

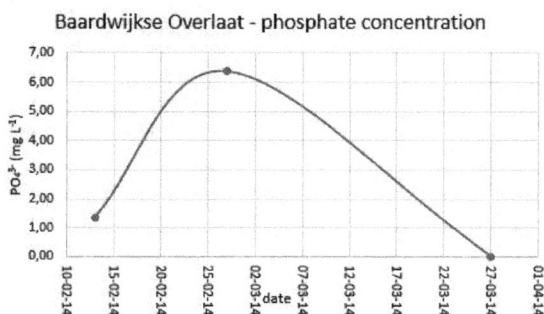

Figure 9

Test samples Elshoutse Wielen

Date	Extinction	PO_4^{3-} (mg L^{-1})
13-02-14	0.067	0.66
27-02-14	0.000	0.00
27-03-14	0.107	1.05
10-04-14		

Table 8

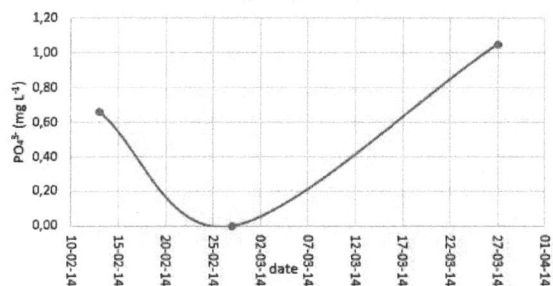

Figure 10

3   Conclusion

At this moment nature is awakening and we just started investigating the biodiversity around the two pools of water. Conclusions are therefore not available at this moment but will be presented in Singapore.

Acknowledgements

First of all we like to thank Raffles Institution and the Water is Life organisation to make this international exchange program possible. We're looking forward to see you in Singapore.

We also like to thank our technical education assistant Sonja en teachers Bernard Klerks and Patrick van Kempen.

References

[1]   Franken, Peter et.al. (2008). *Chemie overal sk vwo deel 2* (3th ed.). Houten: EPN. pp. 112-121

[2]   Beers, Kees & Roo, Ico de (2001). *Water, bron voor onderzoek*, Arnhem: Citogroep

[3]   Natuurmonumenten. *Baardwijkse overlaat.* Viewed Januari 2014, https://www.natuurmonumenten.nl/natuurgebieden/baardwijkse-overlaat

[4]   KNMI. *Klimatologie, daggegevens van het weer in Nederland*, Viewed April 2014, http://www.knmi.nl/klimatologie/daggegevens/selectie.cgi

Water Quality Assessment of the Ruda River as Rybnik Special Resource

Elżbieta Dziubak, Zuzanna Błatoń, Szymon Szeliga

IV Liceum Ogólnokształcące im. Mikołaja Kopernik, Rybnik, Poland,

katarzyna.romaniuk@ivlorybnik.pl, hapy@interia.pl

Abstract

The subject of the study was the Ruda River (Upper Silesia, Poland). The research was based on the physicochemical and biological parameters in order to compare the quality of water in 2002 and 2013. The following physicochemical parameters were evaluated: transparency, pH level, dissolved oxygen, biochemical oxygen demand BZT5 and total phosphorus content. The Belgian Biotic Index (BBI) was used to identify the composition of the benthic macroinvertebrates. The studies were carried out in the spring (March- June) and autumn (September- October) cycles. The analysis of the results showed an improvement in water quality from 3 to 2 classes. The areas around the Ruda River are increasingly used for Recreational purposes.

Keywords

macroinvertebrates, biotic indices, physical and chemical parameters

1. Introduction

Rybnik has a population of 141 thousand inhabitants and is situated in the Silesian Agglomeration in the south of Poland. Rybnik area is located in the basin of the Ruda River and belongs entirely to the Odra basin. The main and biggest body of water in the area is Rybnik Reservoir. It was built to provide and cool process water for Rybnik Power Plant (currently known as EDF Poland). Nowadays the reservoir, also called The Rybnik Sea, is not only an industrial facility but also a place essential from the nature study viewpoint. Moreover, it performs both flood control and recreational functions. [1] One of the most interesting facilities within Rybnik Reservoir is Gzel Reservoir which constitutes a natural habitat for the following characteristic species: yellow water-lily *(Nuphar luteum)*, frogbit *(Hydrocharis morsus)*, water arum *(Calla palustris)*, lesser wintergreen *(Pyrola minor)* and round-leaved wintergreen *(Pyrola rotundifolia)*. [2] Within the town there are some forms of nature conservation connected with hydrogenic habitats, among others the Landscape Park "Cistercian Landscape Compositions of Rudy Wielkie" [3] and ecological areas, eg "Okrzeszyniec". [4]

Rybnik constituted the centre of Rybnik Coal Area and so it was the core of mining and metallurgy industries. Energy, metal and food industries developed as well. Since 1990s, intensive industrial restructuring processes [5] and actions aimed at improving the quality of surface waters have been operating. The main sources polluting the water flowing through the town were municipal sewage, agricultural waste, domestic and industrial waste, as well as wastewater falls and salty mine waters. [1] Rybnik's significant problem was caused by the lack of a sewage system – only the central part of the town (12% of the town surface) contained a sewerage. The lack of a sewage system in the remaining part of the town and insufficient wastewater treatment were the reasons for a big load of pollutants carried into the Ruda River and contaminating both the river and Rybnik Reservoir. This, in turn, had an influence on the poor water quality of the Odra River and finally led to the Baltic Sea pollution.

As a result of implementing the recovery processes mentioned above, the sewage treatment plant, built in 1970-1976 and located in Rybnik-Orzepowice, was expanded and modernized in 2000. [1] Currently it meets the requirements of both the Polish and the European Union Directives. In 2001 the project "Construction of the Sewage System in Rybnik" was started. [6] The expansion of the sewage treatment plant allowed us to obtain the throughput of 27500 m^3/d and the construction of the sewage system enables to connect 150 thousand inhabitants to the target system. The ecological aim of the project was to reduce pollution and improve water quality of the Ruda River, Rybnik Reservoir and the Odra River as well as to protect the Landscape Park "Cistercian Landscape Compositions of Rudy Wielkie."

2. The purpose of investigation.

The purpose of the project is to monitor the condition of the Ruda River. We have been conducting research on the physicochemical and biological (BBI) assessment of water since 2002 when a youth project known as "Cascade" and aimed at monitoring surface waters was started. [7] The basis of the research is formed by the expectations of the improved quality of surface waters resulting from the projects which have been implemented in Rybnik, connected with the expansion of the sewage system. The present report constitutes the comparison of the current river condition to its state in 2002. We aim to show that providing Rybnik with the sewage system and draining wastewater into the existing sewage treatment plant improve water quality and thus increase both the biodiversity and attractiveness of the area for recreational purposes

3. Method

The 50.6-km-long Ruda River is a right tributary of the Odra River and has its source in Żory. It flows through the Cistercian Landscape Compositions of Rudy Wielkie and into the Odra River in the village of Turza. The sampling site was located above Rybnik Reservoir. [Fig.1]

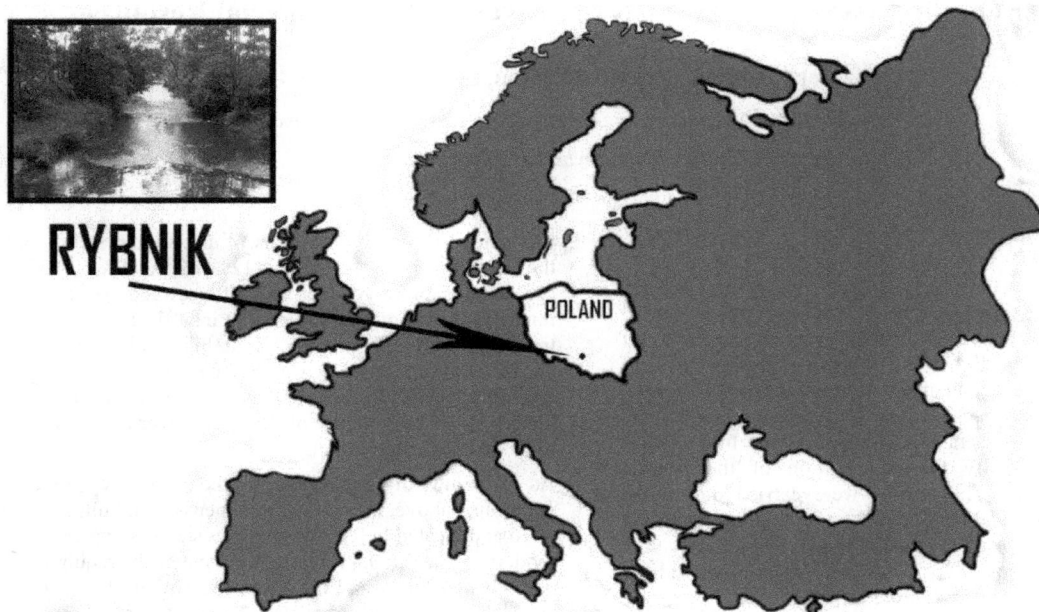

Fig.1 Map of Rybnik reservoir in southern Poland

At this height the river is regulated and has a muddy bottom. The river banks are reinforced and covered with vegetation. The surrounding area contains meadows and reforested areas. There is a cycle path located along the Ruda River. The studies were carried out in the spring (March- June) and autumn (September- October) cycles.

According to the Polish standards, the following physical and chemical parameters were evaluated: transparency, pH level, dissolved oxygen, biochemical oxygen demand BZT5 and total phosphorus content.

The transparency was determined in the field using a Secchi disk suspended with a cord. The white disk is lowered into the water until it is no longer visible. That point is the Secchi disk depth which is the measurement of water transparency [9].

For the determination of phosphorus content a colorimetric method, with ammonium molybdenum and Tin (II) chloride as a reducing agent, was used. Determination of phosphorus in surface waters is very important, because large amounts of this compound get into waters with detergents and their excess causes eutrophication of waters [9].

An electrometric method using a pH meter was used to measure the pH.

The content of dissolved oxygen is one of the most significant indicators of water quality. The smaller content

of dissolved oxygen, the bigger water contamination. To determine the content of dissolved oxygen we used the Winkler titration method [9].

Also the BZT5 indicator, defining the biochemical oxygen demand and indicating organic loading in wastewater, was determined. The dilution method was used [9]. The physicochemical research was carried out in the laboratories belonging to the Silesian University of Technology, the sanitary and epidemiological station, the sewage treatment plant and in our school laboratory.

On the basis of the obtained parameters, the water was classified to the appropriate class of water quality in accordance with the act of the Minister of Environmental Protection from Feb. 11, 2004 (Journal of Laws No 32, item 283 and 284, 2004] [8] [Table1].

Table 1. The admissible values of some physical and chemical parameters of surface waters,
biotic indices: BBI, and related water quality classes.

Water quality indices	Surface water quality classes				
	1	2	3	4	5
Oxygen [mgO2dm-3]	≥ 7	6	5	4	<4
Phosphate [mgPO4dm-3]	$\leq 0,2$	0,4	0,7	1,0	>1,0
[pH]	6,5 – 8,5	6,0 - 8,5	6,0 - 9,0	5,5 – 9,0	< 5,5 , >9,0
BZT5 [O2dm-3]	2	3	6	12	>12
BBI	10-9	8-7	6-5	4-3	2-0
Water quality	very good – very clean waters	Good – clean waters	Satisfactory – slightly contaminated waters	Unsatisfactory – contaminated waters	Poor – highly contaminated waters

The biological studies included collecting macroinvertebrates and the analysis of their qualitative composition. The material was collected interchangeably on both sides of the river over a distance of about 20m, on the same day as the water for physicochemical determinations. The samples were taken with a scoop. Each time 5 samples were collected from the area of approximately 0,25m². The collected species were preserved in alcohol (80%) and the taxonomic identification of macroinvertebrates was done with the usage of appropriate sources [10, 11, 13].

The results of the macroinvertebrates analysis were used for the biological evaluation of the Ruda River water quality. We used the Belgian Biotic Index (BBI).

This index has been serving us for the determination of water quality on the basis of macrobenthos since the "Cascade" programme was started. Although in recent years a number of indices used to assess the biological condition of surface waters have appeared, including the BMWP-PL index, [13,14,15,16], the BBI index has several important advantages for us, namely, it is clear and based on the determination of particular taxa in macrobenthos. Different taxonomic groups should be marked to the level of family or genus in accordance with the guidelines. [Table 2] The biotic index is derived from the Standard Table for Calculation of the Belgian Biotic Index.[Table 3] Having determined the biotic index, we defined the water quality class. [Table 1]

Table 2. Practical limits to identify taxa in the
Belgian Biotic Index [12]

Taxonomic group	Identification level of taxonomic groups
Plathyhelminthes	Genus
Oligochaeta	Family
Hirudinea	Genus
Mollusca	Genus
Crustacea	Family
PLecoptera	Genus
Ephemeroptera	Family
Trichoptera	Family
Odonata	Genus
Megaloptera	Genus
Hemiptera	Genus
Coleoptera	Family
Diptera	Family
Hydracarina	Presence

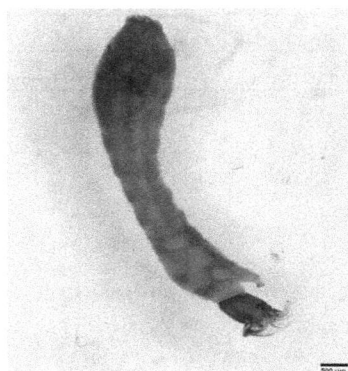

Fig. 2 *Simulidae*

Table 3. Standard table for Calculation of the Belgian Biotic Index [12].

Indicator group	Class frequency	Number of taxa				
		0-1	2-5	6-10	11-15	>16
Plecoptera, Heptagenidae	≥ 2	-	7	8	9	10
	1	5	6	7	8	9
Trichoptera	≥ 2	-	6	7	8	9
	1	5	5	6	7	8
Ancylidae,Acroloxus, Ephemoptera (exc.Heptageniidae)	>2	-	5	6	7	8
	1-2	3	4	5	6	7
Aphelocheirus, Odonata, Gammaridae, Mollusca (exc. Ancylidae, acroloxus, Sphaeridae, Corbicula)	≥ 1	3	4	5	6	7
Asellidae, Hirudinea, Sphaeridae, Hemiptera (exc. Aphelocheirus)	≥ 1	2	3	4	5	-
Tubificidae, Chironomidae thummi-plumosus	≥ 1	1	2	3	-	-
Syrphidae- Eristalinae	≥ 1	0	1	1	-	-

4. Results

The physicochemical parameters of the water in the Ruda River were averaged and presented in Table 4.

The obtained results were compared to the 2002 parameters. [Fig.3]

Table 4. The results of the physicochemical study of the water in the Ruda River in 2013.

Water quality indices	2013			2002
	Minimum value	Maximum value	Average value	Average value
Oxygen [mgO2dm-3]	7,2	12,1	10,5	6,5
Phosphate [mgPO4dm-3]	0,25	0,5	0,3	0,9
[pH]	6,9	7,9	7,1	6,0
BZT5 [O2dm-3]	1	5,5	3,5	9,4
Transparency [m]	0,55	0,65	0,60	0,48
Water class	-----------	----------------	2	3-4

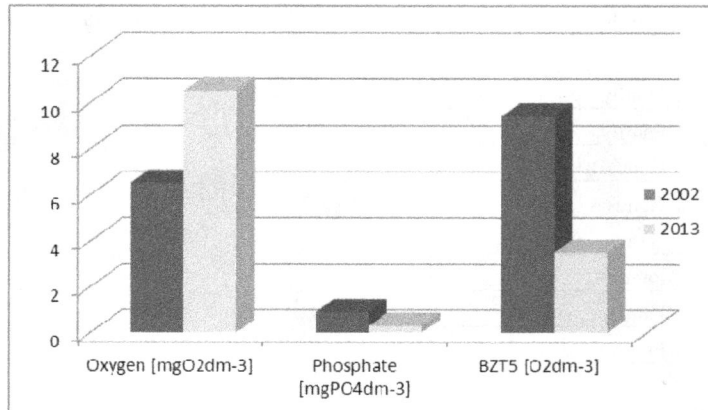

Fig 3. The comparison of the water parameters in the Ruda River during the years 2002-2013

In the Ruda River the presence of 17 taxa was revealed, belonging to Coleoptera, Hemiptera, Mollusca, Hirudinea, Crustacea, Odonata, Diptera, Oligochaeta (Table 5, Fig.3), which served for the determination of the water quality. The majority of taxa identified in the test area was characteristic for moderate clarity of waters and for waters characterized by a muddy or sandy bottom. The comparison of the taxa number to the year 2000 shows a significant rise of taxa (by 11 taxonomic groups) including Odonata and Hemiptera that were not present before. After the results were analysed, it was shown that the Ruda River could be currently classified, according to its biological condition, as being second class water which means the improvement of the water quality by one class compared to 2002.

Table 5 The summary of the designated macrobenthos. The comparison of the results to 2002

Taxon	2013	2002
Coleoptera		
Dytiscidae	+	-
Hemiptera		
Nepa	+	-
Mollusca		
Lymnaea	+	+
Pisidium	+	-
Anodonta	+	-
Hirudinea		
Glossiphonia	+	-
Helobdella	+	+
Eropobdella	+	+
Dina	+	-
Piscicola	+	-
Crustacea		
Asellidae	+	+
Odonata		
Libellula	+	-
Diptera		
Simuliidae	+	-
Culicidae	+	-
Chironomidae	+	-
Oligochaeata		
Tubificidae	+	+
Lumbriculidae	+	+
Taxa number	17	6
BBI index	7	5
Water class	2	3

Conclusion

Our research and observations have shown unambiguously that the water quality of the Ruda River has been gradually improving. Since 2002 oxygenation of the water has increased, there has been a decrease in the phosphate concentration which has led to a significant decrease in the oxygen consumption after 5 days (BZT5 has fallen from 9,4 (2002) to 3,5 (2013). The improvement of the physicochemical parameters is reflected in the biological condition of the Ruda River. There has been an increase in the biodiversity of organisms. The number of taxa groups of macroinvertebrates identified at the sampling site has risen by 11 since 2002.

The results of the measurements on the Ruda River are compatible with the tendency presented in the Report of the Regional Inspectorate for Environmental Protection.
A gradual improvement of water quality has been noticed in the Silesia Province [17].
The obtained results and tendencies to the improving river condition allow to conclude that fewer pollutants, mainly

municipal ones, flow into the regional waters as well as Rybnik Reservoir. It is extremely important in the context of the improvement of the ecosystems qualities as well as the local residents' living standards.

For several years intensive work has been carried out which is aimed at increasing the attractiveness of the area around Rybnik reservoirs:

- Along the Ruda River a recreation path, which is eagerly used by cyclists, runners and walkers, has been created;
- Canoeing events are organised on the Ruda River, which was impossible a few years ago due to the river condition;
- New projects are implemented to develop the areas around the waters for recreational purposes. This year several new investments are supposed to start operating, including the construction of new playgrounds, parks and recreation centres;
- Diving, fishing and sailing clubs are active on Rybnik Reservoir; sailing regattas are organised;
- The areas located near the water reservoirs constitute a significant venue where numerous outdoor photography competitions are held.

Our school, IV Liceum im. Mikolaja Kopernika, also contributes significantly to raising young people's ecological awareness and promoting an active lifestyle. For many years Environmental Awareness Workshops have been organised at our school as well as the annual "Raft" happening. Within the framework of the latter project students from Rybnik secondary schools make rafts from PET bottles. Next the rafts are used in regattas, the idea of which is to provide fun for students and integrate the local community. Finally, the rafts are dismantled and the PET bottles get recycled.

All the ecological and economical activities are designed to make the water areas more attractive. The improvement of water quality changes people's mentality and, as a result, they are more willing to protect the environment and take care of their own physical condition.

References

[1] Environmental Protection Programme for the City of Rybnik – act]

[2] The nature of Upper Silesia, 1997, nr 9, Centrum Dziedzictwa Przyrody Górnego Śląska

[3] Regulation of Katowice Governor No 181/93 from Nov. 23,1993 (Katowice Journal of Laws No 15, item 130).

[4] City Council Resolution No 836/XLIII/2002, the City of Rybnik, from Oct. 4, 2002 (Silesian Journal of Laws No 80/02, item 2898).]

[5] changes in surface water quality of the upper part of the Odra River basin as a result of the restructuring of the coal mining

[6] http://www.rybnik.eu/index.php?id=365.

[7] Kaskada "Cascade" (2003-2010), Śląskie ABC

[8] Ordinance of the Minister of Environmental Protection of 11 Feb. 2004 concerning the classification which is used to present surface and underground waters state, and the ways of monitoring and interpreting the results and presenting the waters state (Journal of Laws No 32, item 283 and 284, 2004)

[9] Hermanowicz W. (1999) Physicochemical Research on Water and Sewage , Arkady, Warszawa, pp.54,159-160, 260-264, 333-334

[10] Rybak J.I.(2000) Freshwater Invertebrates. A guidebook. PWN, Warszawa pp.85

[11] Rybak J.I. (1986) A Guidebook for Identification of Freshwater Invertebrates. PWRiL, Warszawa pp.75

[12] De Pauw N., Ghetti P.F, Manzini P., Spaggiari R. (1992) Biological Assessment Methods for Running Water. In: Ecological Assessment and Control, P.J Newman, M.A. Pivaux, R.A. Sweeting (ed.), Comission of the European Communities, Brussels, pp.217-248

[13] Bis B., (2012) A Guidebook to Assessing the Ecological Condition of Rivers on the Basis of Benthic Macroinvertebrates, NFOŚ, Warszawa pp. 19-26

[14] Gorzel M., Korijów R., (2004) Biological Methods of River Water Quality Assessment, Kosmos, 2, pp.183-191

[15] Kownacki A., Soszka H., Fleituch T., Kudelska D.,(2002) River Biomonitorning and Benthic Invertebrate Communities, Institute of Environmental Protection, Warszawa-Kraków, pp. 71-88

[16] Kułakowski P., Bieniek P., (2003) New Methods of Classification of Surface Waters in Poland. Technical Magazine/ Czasopismo Techniczne z.4-Ś, 200, pp.9-21

[17] The Regional Inspectorate for Environmental Protection, (2013), The State of the Environment in the Silesia Province in 2012, Biblioteka Monitoringu Środowiska, Katowice pp. 58-66

The use of bio-indicators instead of chemical indicators to determine water quality

Roos Peeters, Sophie van Erp and Joost Janssen

Maurick College, the Netherlands, 00801333@maurickcollege.eu

Abstract

Bio-indicators or ecological indicators are taxa or groups of animals that show signs that they are affected by environmental pressure. Bio-indicators that are used to determine water quality are animals that live in water and are caught by active sampling. A special combination of different kind of species in a sample can give an indication of the quality of the water. In this research we will get an insight on the use of bio-indicators. We will compare our results with (1) chemical water determination on the same environment, (2) different seasons and how seasons have an effect on the water quality and (c) data from water samples from different, less or more polluted areas. This project aims to evaluate the differences in water quality measured by the use of bio-indicators and chemical-indicators. We will get an insight on the natural water quality of our own environment and the techniques used to determine this valuable natural monument in our area.

Keywords

Water quality, bio-indicators, chemical-indicators, pollution, organisms, techniques

1. Background information

1.1 Water quality

Water with a good quality is water that is alive. In bio-diversity there are plants and organisms which can be divided in positive and negative species for the development of a good water quality (M. Scheepens, 2014).

The evacuation of heavy metals in fresh water often leads to very long-term damage (P.C. Brooks, 1995). The water of rivers and lakes keeps itself clean by tiny animals and plants, which break down substances by chemical processes. These substances have a "natural" organic source, such as oil (Peter Strauss, 1998). Water cannot clean itself in the case of contamination by heavy metals and other non-natural, inorganic substances such as copper, zinc and cadmium. In addition, the discharge of cooling water often causes thermal pollution, pollution by heat. Whenever hot water is discharged, it is more difficult for the organisms that live in the water to assimilate waste materials (M. de Jonge, 2011).

The pollution which the aquatic flora and fauna cannot process descends to the bottom of the water or adheres to the sludge particles. Frequently, the harmful substances translocate in slow-flowing mud whereby they impair the animal life over a wide area. As well, the contaminated water urges into groundwater and, eventually, to agricultural land.

1.2 The physical-chemical determination

Nowadays there is a whole range of techniques and methods for determining water quality based on physic-chemical factors (Morin, 2006). You can measure the acidity
(pH), electrical conductivity, turbidity and temperature. Also, the amount of available oxygen, nutrients (N, P, Na, K) and heavy metals are measurements that will be charged. The physic-chemical conditions are spot samplings and, therefore, provide merely a snapshot of the water. Since these variables vary greatly as a function of space and time, they require a large number of measurements to arrive at an accurate assessment of the quality of water. (De Pauw & Vanhooren, 1983; Metcalfe, 1989).

1.3 Bio-indicators

A bio-indicator is an organism or group of organisms that allow to characterize the state of an ecosystem based on biochemical, cytological, physiological, ethological or ecological variables. This definition includes different characteristics of organisms, populations or communities influenced by the environmental impact (P. Blandin, 1986). Bio-indicators do not only reflect changes in the environment, but often say something about the intensity of exposure to a particular substance. This is done through a measurable response to both physiological, biochemical and on a behaviour biological level (M. de Jonge, 2011).

2. Introduction

Macro fauna includes small invertebrates that can be divided in a variety of taxonomic groups. In the identification of the macro fauna, the fact that the species' composition is not the same throughout the year needs to be taken into account. Some species exist all seasons, while others merely live in the spring and summer (Bakkers, S.; Bergenhenegouwen, R.; Bloemberg, M.; Broecke, S. van den., 2012). That is why we have decided to do our measuring in 3 different seasons: autumn, winter and spring. As the taxonomic groups each impose different requirements on their environment, they can be used as bio-indicators. The presence of groups that place high demands on the water quality, denotes to a high water quality. The absence of these groups can denote to a poor water quality, however, that is not necessarily true (Metcalfe, 1989).

It is provided that the quality of water can be determined by using chemical indicators (Morin, 2006). They monitor the pH-value, dissolved salts and water temperature. We investigate whether it is possible to get the same conclusions by the use of bio-indicators.

The existence of taxonomic groups is also dependent on a number of (a)biotic factors. The type of composition and the population size of macro fauna depends on the amount and different types of plants in the vicinity of water. Animals use these plants as a source of food or shelter from predators such as fish and amphibians (Bakkers, S.; Bergenhenegouwen, R.; Bloemberg, M.; Broecke, S. van den., 2012). Which plants are known to occur and in what quantities is dependent on the amount of dissolved nutrients in the water, and other factors such as pH, soil type and pollution. Another important factor is the water depth and light availability (Descy & Micha, 1988).

Plants need light to grow. In deep water, light never penetrates all the way through to the bottom, because plankton absorbs the light. Animals that eat green plant parts at the bottom, are also likely to be absent. A high diversity of organisms can be found in shallow shore zones in particular (Bakkers, S.; Bergenhenegouwen, R.; Bloemberg, M.; Broecke, S. van den., 2012). Also the acidity determines the occurrence of organisms. Some species like acidic to slightly acidic water (pH 4-6), other types hold true to more basic waters (pH>7). In general, the more acidic the water, the more difficult it is for plants and animals to survive (Bionaturalis, 2013).

Other important key factors are currents, salinity and oxygen. Also temperature, presence of food and concentrations of nitrate and ammonia are important factors (Higler, B. (2006)), (Greenhalgh, M. & Ovenden, D. (2007, vert. uit het Engels, cop. 2010)), (Cuppen, J. & Scheffer, M. (2005)). In our research, we look at the pH, NO3 and O2 value, temperature and turbidity. In this way we gather the quality of the water and then we examine the results. It can be compared with our own bio-indicators. The differences between landscapes will affect our measurement results. Therefore, we will measure in 3 different places. 1) Where the river flows through the city, 2) where the river flows through meadows and 3) where the river flows through a forest. A second factor is the influence of different seasons. For example in the winter there are other living organisms and plants than in the summer. Therefore we measure in the fall, winter and spring. We couldn't do our measurements in the summer because we started our project in the fall.

3. The purpose of the investigation

Research question
Is the use of bio-indicators a good replacement for the use of chemical-indicators to measure the water quality?

Research hypothesis
The use of bio-indicators is a good replacement for the use of chemical-indicators to measure water quality.

3.1 Method of the investigation

Materials

bio-indicators	Chemical- indicators
nets	Stainless steel temperature probe
Identification tables	Dissolved oxygen probe
pots	Turbidity sensor

Tubes	PH-sensor
Microscope	Water samples
Pincers	beakers
Pots	flasks
Tables	Cuvette (cuvette length: 1 cm)
Microscope	Pipette
	NO3 Test (Sera)
	Accelerometer
	1000 mg /l NO3 (standard)
	Spectrophotometer PU8620 (540 nm)
	Lab Quest Vernier

Method
In our water project we have chosen to take water samples from the Dommel, a river that flows nearby Den Bosch. To get a good idea of the water quality of the river, we did various measurements in autumn, winter and spring. We did our measurements on the banks of the river, because the most plants and, therefore, also the most organisms are there. We have done all the measurements three times, to get an average of the values. In order to obtain the best representative macro invertebrates possible, the samples are taken in the vicinity of both plants and the bottom. We determined the organisms we found at the different locations. Mostly, we investigated our chemical measurements in the lab.

3.2 Chemical indication

We needed different types of equipment to measure the water quality of the Dommel. Every device has its own function, like the determination of oxygen level. We borrowed five various kinds of equipment from our school. We worked with an oxygen sensor, pH-sensor, thermometer, turbidity meter, and we have measured the nitrate by using a spectrometer. There are different substances stored in the water which we can measure with our equipment.

Moreover, clear water is vital for many plants and organisms.

At the three locations (the forest, agriculture and woods), we have taken water samples from the bottom and surface. We chose to do this because the substances that are stored in the bottom water can be different from the substances that are stored in the surface water. We did this in the fall, winter and spring in order to measure the differences between the seasons.

3.2.1 Meaning of chemical measurement

Temperature
The temperature of a waterbody directly affects many physical, biological and chemical characteristics.

Temperature directly affects the metabolic rate of plants and animals. Aquatic species have evolved to live in water of specific temperatures. If the water becomes colder or warmer, the organisms do not function as effectively, and become more susceptible to toxic waste, parasites and diseases. Changes in the average long-term temperature may cause differences in the species that are present in the ecosystem. So a constant temperature is ideal (Waterwatch Australia, 2002).

Oxygen

Oxygen is essential for almost all forms of life. Aquatic animals, plants and most bacteria need it for respiration (getting energy from food), as well as for some chemical reactions. In some circumstances, water can contain too much oxygen and is said to be supersaturated with oxygen. In this supersaturated environment, the oxygen concentration in fish's blood rises. When the fish swim out into water that has less dissolved oxygen, bubbles of oxygen quickly form in their blood, harming the circulation (Pondlibary 2003). The oxygen level is good/fair when the saturation percentage is between 50% and 70%. The oxygen level is considered excellent when the saturation percentage is above 70% and under 100%. Between 100% and 120% the water quality is considered good. If the saturation gets more than 120%, it may indicate that there are a lot of algae, which is bad for the water quality (Onecuesystems 2013).

PH-value

Values of pH range from 0 (highly acidic) to 14 (highly alkaline). Where water has no net alkalinity or acidity it is said to be neutral and has a pH of 7. Many compounds are more soluble in acidic waters than in neutral or alkaline waters. The pH of the wet area around roots affects nutrient uptake by the plants; pH also affects the solubility of heavy metals in water and the concentrations of total dissolved solids in rivers. All animals and plants are adapted to specific pH ranges, generally between 6.5 and 8.0. If the pH of a waterway or waterbody is outside the normal range for an organism it can cause stress or even death to that organism (state of environment report 1994). To be more precise there is a scale to classify different levels of pH values, see table 1 (appendix).

Nitrate (NO3)

The most common nitrogen compounds are ammonia (NH3), nitrate (NO3) and nitrite (NO2). They occur in dissolved, particulate and gaseous forms.

- As nitrate (NO3) is soluble and easily absorbed by aquatic organisms, it is the most meaningful form for Waterwatchers to test. (table 2, appendix) (Bioplek 2013)
- Ammonia (NH3) is a product of the decomposition of organic waste and can be used as an indicator of the amount of organic matter in the waterway.
- Nitrite (NO2) is toxic to humans and other animals.

Nitrate compounds can be found in surface waters and in groundwater. The element nitrogen is continually recycled by plants and animals (Waterwatch Australia 2002).

Turbidity

In general, the more material suspended in water, the greater the water's turbidity and the lower its clarity becomes. Suspended material can be particles of clay, silt, sand, algae, plankton, micro-organisms and other substances. Turbidity affects how deep light can penetrate into the water. Turbidity can indicate the presence of sediment that has come from construction, agricultural practices etcetera. Suspended particles absorb heat, so water temperature rises faster in turbid water than it does in clear water. If penetration of light into the water is restricted, photosynthesis of green plants in the water is also restricted. This means less food and oxygen is available for aquatic animals. Suspended silt particles eventually settle into the space between the gravel and rocks on the bed of a waterbody and decrease the amount and type of habitat available for creatures that live in those crevices. Suspended particles can clog fish gills, inducing disease, slower growth and, in extreme cases, death (Waterwatch Australia 2002). The scale of pollution caused by turbidity is shown in table 3 (appendix) (Themes 21, 2011).

3.3 Bio-indication

With a biological quality assessment, we can examine what extent we are dealing with, clean or dirty water. On the basis of small aquatic creatures contained in the water, known as macro fauna, the quality of the water can be determined fairly accurately. With increasing pollution macro invertebrates (invertebrates that are visible to the naked eye) disappear, causing the diversity of species to decrease. This has to do with the discharge of the polluted water into the surface water. (Betavak, 2006) There are fewer species of organisms and the number of individuals of each species is bigger. We caught different organisms using fishing nets and determined them using identification tables.

Furthermore, any plant comes into its own in its environment. On one side of the water you can find for example, less or no plants of a certain type, while you can find the plant on the other side of the water. Plants that grow on the shore side or in the water, also give an indication of water quality. We also identified the plants and looked at what impact the plants have on the water quality.

By taking water samples and capturing organisms, we looked at what organisms do to maintain the water quality as effective as possible. We also looked at the environmental conditions. In non-polluted water, the common types of animals and the number of individuals of each species is (relatively) small (large diversity). If the water is more seriously polluted (disturbed), the diversity is lower. There are fewer species of animals and the number of individuals of each species is greater. For example, an organism might live in highly polluted water such as worms, or in pure water (stone fly). They can give an indication of water quality.

Using table 4, 5 and 6 (APPENDIX), we can determine, with our found organisms, whether the water is very dirty or pure. When we find an organism, we use table 4 or 5 to see what the species means for the water quality. Using table 6, we can see directly whether the species is has a positive or negative influence on the water quality.

4. Results

Bio-indication

Table 7

Environment in spring	found aquatic animals and aquatic plants
agriculture	Dace, chub, shellfish, caddis fly , water isopods , boatswain, larva of a dragonfly, skater, worms, star duckweed, groin grass
city	Groin grass, Cabomba, pondweed, worms, skater
forest	Groin grass, star duckweed, pondweed, amphipods, skater, shellfish

Table 8

Environment in winter	Found aquatic animals and aquatic plants
agriculture	Boatswain, caddis fly, skater, star duckweed, pondweed, chub
City	Skater, groin grass, cabomba, pondweed
forest	Groin grass, star duckweed, pondweed, amphipods, skaters

Table 9

Environment in autumn	Found aquatic animals and aquatic plants
Agriculture	Skaters, boatswain, star duckweed, waterisopod , groin grass, shellfish
City	Skaters, worms, groin grass, Cabomba, pondweed
Forest	Amphipods, skaters, shellfish, star duckweed, pondweed, Vlokkreeftjes, waterlopers, schelpdiertjes, sterrenkroos, waterpest, liesgras, frog

In the tables above (table 7, 8 and 9) is indicated which organisms we found on which place.

Agriculture
In the spring, we found a larva of a dragonfly (Libellus) and a dace (Sinilabeo), this means that the water has a good oxygen-value. We also found shellfish (Gastropoda), Star duckweed (Calliriche Platycarpa) and Groin grass (Glyceria Maxima Variegata). These organisms only live in nutritious water. Only a few plants and organisms are able to live in nutritious water. The fact that we've found worms (Lumbricus) and water isopods (Asellul Aquaticus) indicates that the water is highly polluted. In contrast, the life chub (Leucisus Cephalus), caddis flies (Trichoptera) and larvae of the dragonfly only live in good quality water. Therefore, the found organisms contradict each other pretty disappointingly. An explanation for this could be that we did our measurements on a pier with boats. This is a place where many people come and they also throw trash into the water. This attracts worms and water isopods and so the oxygen-value can still be good. During the winter, the water seems to be highly polluted. We have found star duckweed and pondweed (Eloda Sp.), both denote a moderate water quality. We also found a chub, caddis fly, boatswain (Notonecta Glauca) and skater (Gerris Lacustris). On the other hand, these organisms indicate that the water quality is good. In the winter, the water quality is better than in the spring. This could be because the usage of boats in the winter is lower, so less people go to the pier. This causes the water to be less polluted by human influence. During the fall we found the largest number of organisms that indicate contaminated water. We found star duckweed, groin grass and water isopods. We found a few stone flies (Plecoptera), boatmen, shellfish and skaters, but these were relatively small. The water quality during the fall was bad, because the water was very nutritious. Waters may be lost because they get overgrown in relation to the excess of nutrients that end up in the environment.

City

In spring and fall, we found the skater, worms, groin grass, pondweed and cabomba (Cabomba Caroliniana). This indicates a very moderate water quality. The skater normally lives in water with a good quality, but the presence of the worm, groin grass, pondweed and cabomba doesn't indicate a good water quality. The plants live in nutrient-rich water and the worm lives of the waste in the water. In the winter we found the same organisms, except for the skater. This means that the water in the city in the winter, according to the bio-indicators is heavily contaminated. The worm lives of the waste that is in the water in the city and 3 plants all live in nutrient-rich water. The Cabomba could be able to overgrow across the waterway, for this he needs nutrient-rich soil. Only few plants and animals can live in nutrient-rich waters, so this is bad for the water quality.

Woods

In the spring we found shellfish, amphipods (Dikerogammarus villosus), skaters, star duckweed, groin grass and pondweed in the woords. This is a wide variety of aquatic organisms. The flea crustaceans stand together with the skaters and shellfish for excellent water quality. The star duckweed, groin grass and pondweed have a negative effect on the water quality. The amphipods can live well in the forest, because they live on soft ground in very cold oxygenated water. In this case, the water in the forest flowed over a smooth sandy bottom. In the winter we found the same in the woods except for the shellfish. It seems that

the water quality result was worse, but it is possible that shellfish does not survive the winter. In the fall, we found the same organisms as in the spring, only the frogs were still there. For frogs to be able to live, the water quality should be good. The water quality of the forest is fairly constant throughout the year. This is probably because a relatively small amount of people goes to the water in the forest. Thus, there is little, to no human pollution of the water. This is a big contrast to the water quality in the city.

In an agricultural environment, we found a high diversity of organisms throughout the seasons. However, in the city, we found a low diversity of insects. In contrast, we found more different species of aquatic plants in the city, than anywhere else. The low diversity of organisms is probably because there is sludge on the bottom of the water. Through the seasons, the number of organisms changes. Moreover, there are fewer organisms in the winter than in the spring, because organisms depend on certain temperatures. Another thing causing a change in the results is whether we measure on a cloudy or on a sunny day. On a sunny day, we obviously find more bugs than on a cloudy day, because on a sunny day, animals go to the surface.

Chemical-indication

Graphic of the river bottom in spring

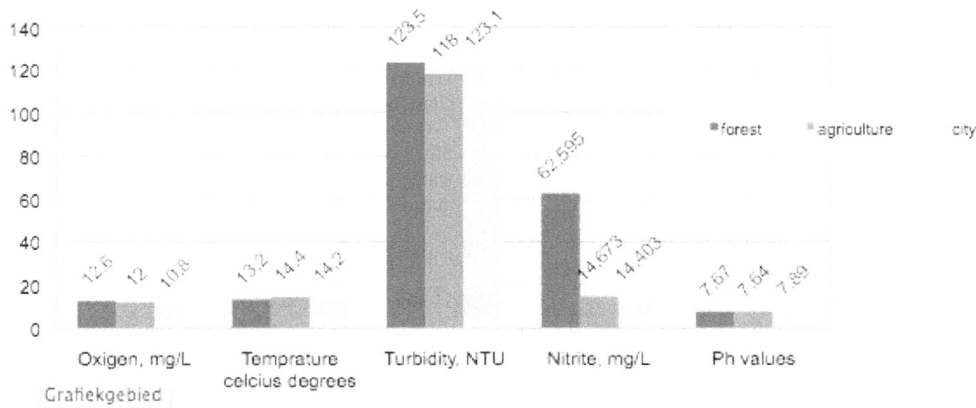

Figure 1

Overall most of the pH values are good. And everywhere the water is turbid. For the forest in spring the oxygen levels are good and the nitrate levels differ from a little polluted to serious polluted. Overall the water quality is good. In the winter the oxygen levels are good to excellent, the water is moderately to not polluted at all. Nitrate and pH-levels are excellent here, which indicates a great water quality. In autumn the oxygen levels are also excellent, the nitrate pollution is moderate but the pH is excellent. This indicates a good water quality.

When we look at the agricultural location in the spring, we notice that the oxygen levels differ from good to bad. The nitrate levels are moderately polluted. This indicates that the water quality is ok, not good but also not bad. In the winter the oxygen goes from good to excellent, nitrate level

is moderate but pH levels differ from good to excellent! This indicates a relatively good water quality.

In autumn the oxygen level goes from good to not so good. But on the other hand the nitrate is just a little polluted. The water quality was equal in winter.

And finally, we tested the city. In spring the oxygen differs from good to excellent and the nitrate levels are moderate. With this taken into count, the water quality turns out to be ok. In the winter the oxygen is good, the nitrate values differ from a little polluted to seriously polluted. PH differs from good to excellent. The overall water quality is good. During the autumn we notice that oxygen levels differ from good to not so good. And nitrate is a little polluted. That means the water quality is ok.

5. Conclusion

Comparing the results we have found, we come to a clearer conclusion with our bio-indicators than with our chemical indicators. The bio-indicators obviously show that the water quality in the forest is the best, which is the most logical. When we take a look at our chemical-indicators, this is not the case. We are not able to make a clear conclusion about the water quality looking at the chemical-indicators. A reason for this could be that we do not have enough chemical measurements. It becomes apparent that for students, children and maybe even specialists, it would be easier to use bio-indicators instead of chemical indicators. The use is easier and possibly in some cases even more reliable: Think of measurement uncertainties and errors in the measurement, in addition the fact that it is easier to measure bio-indicators than chemical indicators.

In the chapter "water quality", we state that living water is the definition of a good water quality. Which suggests that research based on bio-indicators is better than research based on chemical-indicators. Because they directly tell us something about the quality of the water, while the chemical-indicators tell us something about the conditions of the water. The water conditions for life in water go from great to terrible. After collecting a lot of data about the chemical circumstances in the water, a good prediction whether life will flourish or not can be made. However, you will never be certain. So bio-indicators are more direct because they are the water quality, instead of the indicators of it. Moreover, with the use of chemical-indicators, there is a greater chance of measure uncertainties. Also in our own study. The Oxygen values are highly remarkable, they are so high we doubt the quality of our measuring equipment. Also the turbidity-value draws our attention.

Our hypothesis was "To measure the water quality, the usage of bio-indicators is more reliable than the usage of chemical-indicators." After looking at our results, we proved ourselves to be right. Our experiment shows that it is possible to use bio-indicators instead of chemical indicators.

One of the goals of the research was to form a foundation for further study. So in the future it is possible to go further into the details and cross the boundaries of our research. During our research we learned very much. We were able to process a lot of what we learned in our study, but not all. Something to keep in mind is that when you take a sample you can keep it for 24 hours, after 24 hours it is unreliable.

Another thing is, when you choose what you are going to measure it might be smart to consider phosphorus, it is a very important substance in the water cycle. Furthermore, we advise to test all of your measuring equipment on reliability, so you will not end up with wrong measurement values. Additionally, always measure on the same time. Both chemical values as bio findings can change over even a small amount of time. Don't forget to check the weather either, because insects will be harder to find on a cloudy day than on a sunny day. And even when the conditions for measuring are ideal you have to know where the animals hide. Finally, if you have the chance to measure in four different seasons, take that chance!

References

P.C. Brooks, *Metal residues in soils previously treated with sewage-sludge and their effects on growth and nitrogen fixation by blue-green algae,* 1986

Maarten de Jonge, *Respons van aquatische organismen op metaalverontreiniging in natuurlijke waterlopen: vergelijking diatomeeën en macro-invertebraten,* 2006

Bakkers, S.; Bergenhenegouwen, R.; Bloemberg, M.; Broecke, S. van den. *Waterkwaliteit & aquatische macrofauna in 't Zwanenbroekje, 2012*

P. Maesen M. Sanne I. Van Avermaet *BioNaturalis 4* (biologie leerwerkboek), 2013

Waterwatch Aurstralia national technical manual, Physical and Chemical Parameters.

State of environment report – Victoria's inland waters in water watch Victoria, a community water monitoring manual for Victoria 1994

Bioplek, chemical quality analysis

Thames 21, A water quality analysis of the River Lee and major tributaries within the perimeter of the M25, from Waltham Abbey to Bow Locks, 2005

Peter strauss, *natural attenuation of organic compounds,* 1998

APPENDIX

Table 1. Effects of pH Levels on Aquatic Life

pH-value	Effect
3.0 – 3.5	Fish aren't likely to survive for more than a few hours in waters with this pH-value, although some plants and invertebrates can be found at pH levels this low.
3.5 – 4.0	Known to be lethal to salmonids.

4.0 – 4.5	All fish, most frogs, insects absent.
4.5 – 5.0	Mayfly and many other insects absent. Most fish eggs will not hatch.
5.0 – 5.5	Bottom-dwelling bacteria (decomposers) begin to die. Leaf litter and detritus begin to accumulate, locking up essential nutrients and interrupting chemical cycling. Plankton begin to disappear. Snails and clams absent. Mats of fungi begin to replace bacteria in the substrate. Metals (aluminum, lead) normally trapped in sediments are released into the acidified water in forms toxic to aquatic life.
6.0 – 6.5	Freshwater shrimp absent. Unlikely to be directly harmful to fish unless loose carbon dioxide is high (in excess of 100 mg/L).
6.5 – 8.2	Optimal for most organisms
8.2 – 9.0	Unlikely to be directly harmful to fish, but indirect effects occur at this level due to chemical changes in the water.
9.0 – 10.5	Likely to be harmful to salmonids and perch if present for long periods.
10.5 – 11.0	Rapidly lethal to salmonids. Prolonged exposure is lethal to carp, perch.
11.0 – 11	Rapidly lethal to all species of fish.

Table 2. Nitrate-value and water quality

Pollution by nitrate	no-pollution	Slightly polluted	Moderately polluted	Seriously polluted	Strong polluted
Mg/L	< 4	04 - 12	12 - 36	36 – 108	> 108

Table 3. Turbidity and water quality

Pollution by turbidity	clear	intermediate	Medium turbidity	turbid	Very turbid
NTU	<3	3 - 29	29 - 88	88 - 258	> 258

Table 4. The found organisms

organism	Function of the organism
Calopteryx splendens	A larva ensures a reasonable quality of the water. This organism is found only in water, with a relatively high oxygen saturation. A rugged, overhanging riparian vegetation is important as a vehicle seat.
Amphipods *Dikerogammarus villosus*	Amphipods are useful animals, they live at the bottom of waste and graze vegetation, searching for fouling, algae on the leaves of aquatic plants. They also eat garbage on the bottom.
Caddis-fly *Trichoptera*	The larvae eat the leaves that fall into the stream, which prevents the ditch silts. This may improve the quality of a stream. The abdomen of the animal is in a tube of grains of sand.
Worms *Lumbricus*	Worms often live in highly polluted water between plants, on the ground or on the shore.

Chub	Chub, like barbel and nase, need variety in the river, so gravel and sandbars and shallow sheltered areas with riparian vegetation and underwater plants. By channeling, eutrophication and chemical pollution of major rivers such habitat may disappear. Furthermore, the oxygen content of the water should not fall too much.
Leuciscus Cephalus	
Dace	The dace is relatively insensitive to contamination, but gets into trouble at low oxygen concentrations (less than 7 mg / l) at all stages.
Sinilabeo	
Larva of a dragonfly	Dragonfly larvae needs oxygenated water. Most species can live only in relatively clean water.
Libellus	
Boatswain	Boatswain are predatory insects. He or she attacks everything that moves. This organism is a positive species which does not damage the quality of the water.
Notonecta Glauca	
Stonefly	Stoneflies are very sensitive to polluted water, they are known as indicator species for good water quality.
Plecoptera	
Skater	Skaters are found in non-polluted water. With soap contaminated water does reduce the surface tension, allowing the skater, he falls through the surface and drowns.
Gerris Lacustris	
Watersnails and shellfish	Water Snails and other shellfish are very vulnerable when it comes to the presence of metals in the water. For the formation of their house, the slag also high hardness and good acidity on price. Snails found especially in rather slowly flowing, nutrient-rich water with many plants.
Gastropoda	
Water isopod	The isopod lives of organic material on the bottom of the water, and is a typical scavenger that can occur in large amounts. The woodlice live in stagnant water, often contaminated water at the bottom and along the shoreline between dead plant parts
Asellul Aquaticus	

Table 5

waterplant	function
Star duckweed	This waterplant needs sunny to slightly shaded, moderately nutrient-rich water. But not strong fertilized, clear, neutral, shallow, stagnant to flowing, fresh water with a mineral to organic soil.
Callitriche Platycarpa	
Groin grass	This plant indicates a nutrient-rich water. The water plant is found on nutrient-rich soils.
Glyceria Maxima 'Variegata'	
Pondweed	Waterweed is an oxygen plant. It is very useful for streams and rivers, it is eaten by some include fish and crayfish.
Elodea Sp.	
Cabomba	The plant requires a fertile ground. The plant is able to overgrow, causing problems for recreation and threatens biodiversity.
Cabomba Caroliniana	

Table 6

Animal species	Water quality
No water creatures	Deadly contamination
Bloodworms, rat tail grub	Heavily polluted water
Water isopod, pendulum worms, bloodworms and a rat tail grub	Heavily contaminated
amphipods, water isopod, pendulum worms, bloodworms and a rat tail grub	Contaminated
Larva of a caddis fly, amphipod, water isopod, pendulum worms, bloodworms and a rat tail grub	Moderately polluted
Larva of a mayfly, Larva of a caddis fly, amphipod, water isopod, pendulum worms, bloodworms and a rat tail grub	Fairly pure water
Larva of a stonefly, Larva of a mayfly, Larva of a caddis fly, amphipod, water isopod, pendulum worms, bloodworms and a rat tail grub	Very pure water

Locations

Graphics of the chemical measurement

Woods surface

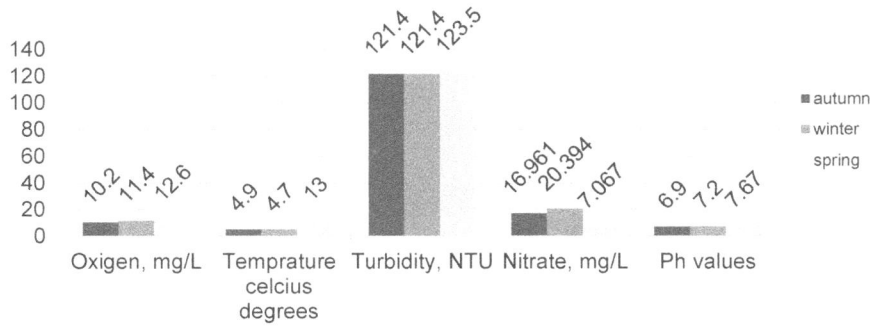

Woods bottom

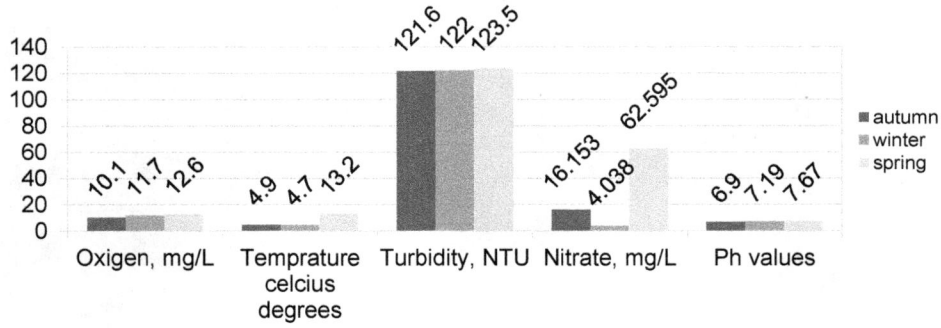

Agriculture surface

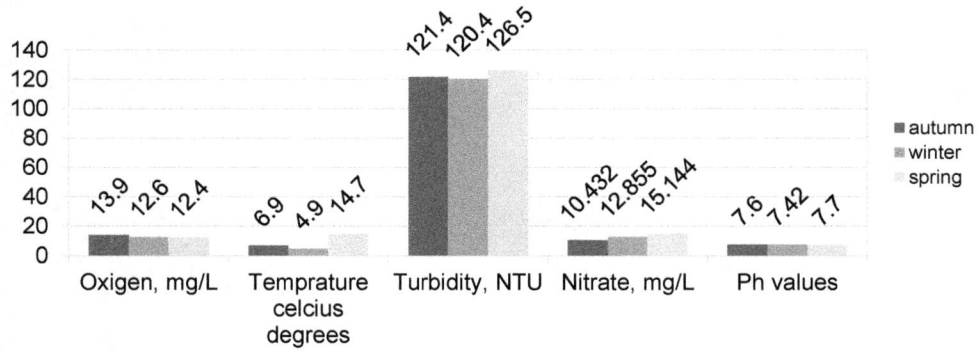

Agriculture

bottom

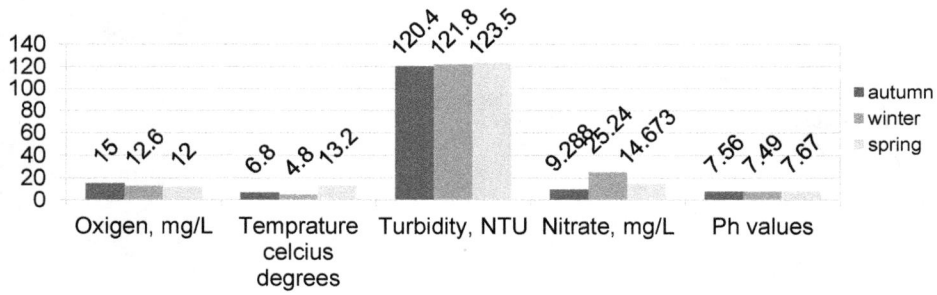

City surface

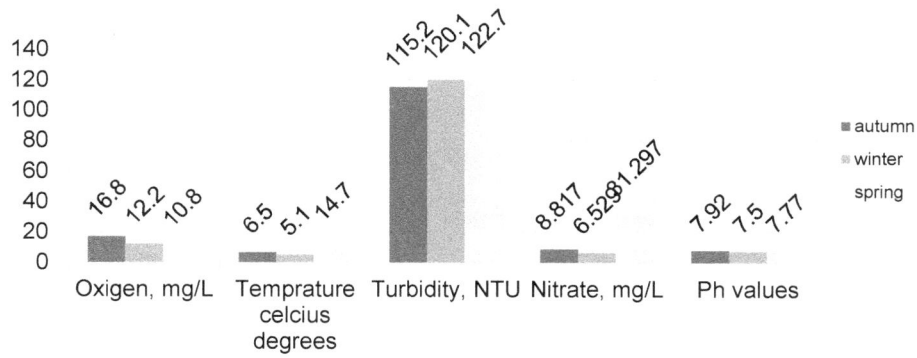

City bottom

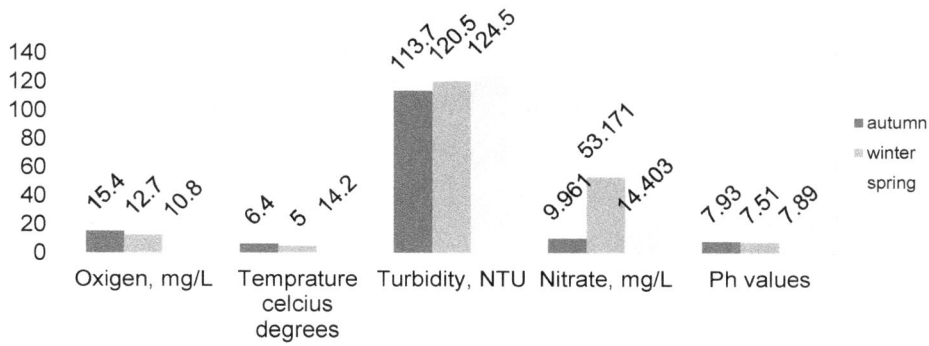

Autumn   surface

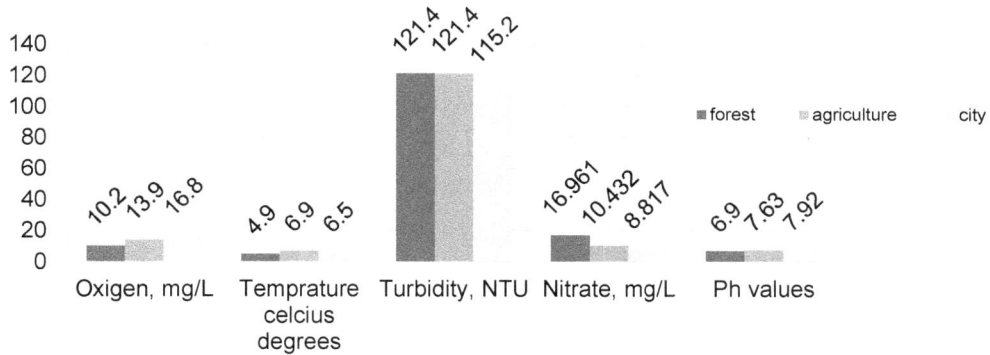

Autumn bottom

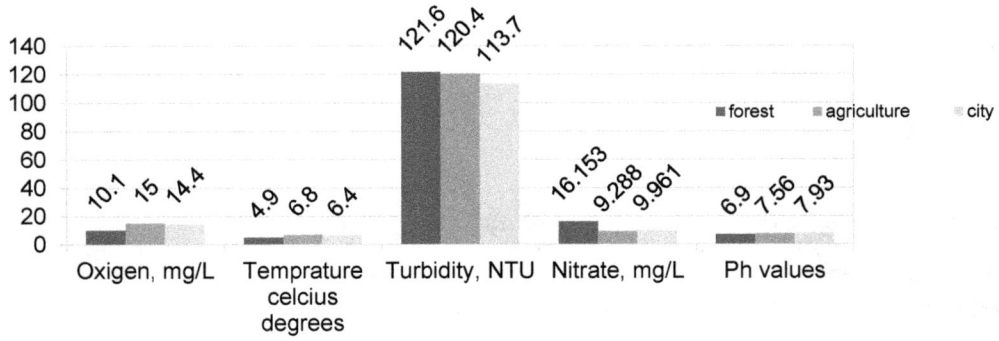

Winter surface

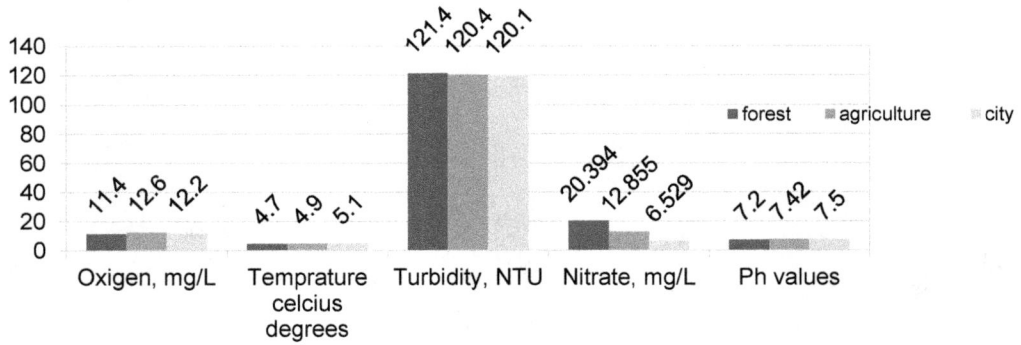

Winter bottom

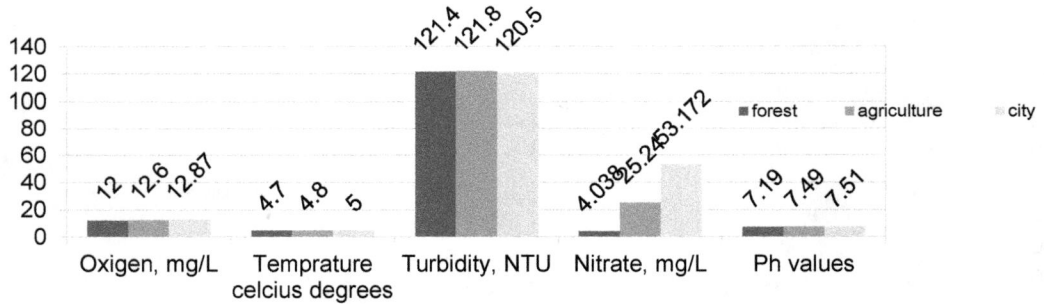

Spring surface

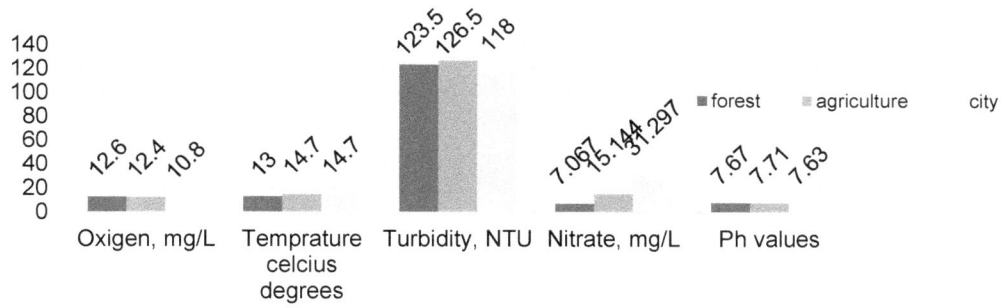

Spring bottom

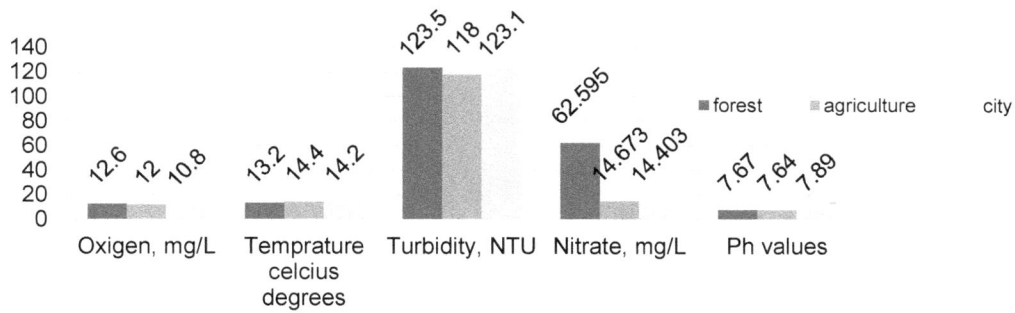

Calibration

ABS	mg/l
0	0
0,097	10
0,443	25
0,705	50

autumn forest	surface	soil
ABS	0,252	0,24
mg/l	16,961112	16,15344

autumn agriculture	surface	soil
ABS	0,155	0,138
mg/l	10,43243	9,288228

autumn city	surface	soil
ABS	0,131	0,148
mg/l	8,817086	9,961288

winter forest	surface	soil
ABS	0,303	0,06
mg/l	20,393718	4,03836

winter agriculture	surface	soil
ABS	0,191	0,375
mg/l	12,855446	25,23975

winter city	surface	soil
ABS	0,097	0,79
mg/l	6,528682	53,17174

spring forest	surface	soil
ABS	0,105	0,93
mg/l	7,06713	62,59458

spring agriculture	surface	soil
ABS	0,225	0,218
mg/l	15,14385	14,672708

spring city	surface	soil
ABS	0,465	0,214
mg/l	31,29729	14,403484

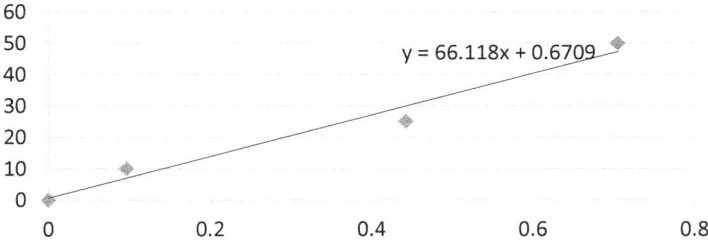

calibration NO3

♦ mg/l ——— Linear (mg/l)

$y = 66.118x + 0.6709$

Water versus brown coal: Tremendous elimination of water resources caused by open cast mining

Stefanie Schiller, Lara Franzmann, Jan-Philip Benthien

Gymnasium Rheindahlen, Germany, kuehnmg@t- online.de

Abstract

The second largest brown coal open cast of the world, named "Garzweiler II", is located in the German Lower Rhine Area, near the town of Cologne. Aiming for the generation of electrical power, the winning of brown coal causes the wasting of an amount of finest groundwater, that millions of people could live from for decades. But furthermore, in an enormous area offside the open cast, the tremendous groundwater lowering injures and endangers ecological very valuable wetlands and endangers the quality of water and the whole local water economy. Additionally, the groundwater lowering causes mining damages of buildings, roads and other infrastructural elements. After the mining ends - around the year 2045 - it will take 40 years, until the residual lake is filled with water. But the quality of this water and the whole future water economy of the region is endangered by acidification of water caused by winning processes. We will describe all these effects caused by brown coal mining.

Keywords

Groundwater, drinking water, wetlands, brown coal

1. Introduction

1.1. That's our native country & school

The Gymnasium Rheindahlen is situated in the west of the town of Mönchengladbach in Northrheinwestfalia near by the border to the Netherlands. Our city has 255.087 inhabitants and is located near cologne.

There are nearly 650 students at our school, starting at the age of 10 up to 18 years and the exam is the German "Abitur". There are three parallel classes a year, 28-32 students in each class. Our school is an elite school of sports and has a focus on international exchanges.

Figure 1: Map of Germany – Mönchengladbach

Figure 2: school building

1.2. Our intention

We are interested in the water problems caused by the brown coal mining, because next to our town there is a large area for the extraction of brown coal, what means, that there are immense consequences for our environment, population & future. In the following presentation, we are going to explain these consequences in hope to find solutions and to make this problem become popular, for the government to see that they should stop this kind of generation of energy.

2. Brown coal

2.1. What is brown coal mining?

Brown coal is a brown – black sedimentary rock and it is mainly used for the generation of energy.

Next to our city, the second largest brown coal open cast mining cast of the world, named "Garzweiler" is located. Aiming for the generation of energy of electrical power, the winning of brown coal causes the wasting of an amount of finest groundwater, so that millions of people could live from for decades.

2.2 How did brown coal develop?

The development of brown coal in Middle Europe, which was, back then, defined by marshy and moory landscape, began approximately 20 million years ago. At this time there was still subtropical climate.

Trees, bushes, ferns and grasses fell, when they died, into the muddy water, however they could not humify since they were provided an airtight seal by the water of the moors and marshes. Firstly microorganisms subverted the dead plants to turf. On top of the new built layer of turf new plants could grow. This cycle of growing, dying and sinking in wet underground repeated again and again caused by the gradual ground's lowering over millions of years.

As a result the layers of turf became thicker and heavier. Over time the climate changed and a sea covered the area and deposited a thick layer of sand. The heavy layer compressed the underlying layers more and more and because of the pressure the turf became brown coal.

Today the ocean's and the river's depositions of "Rhein-Maas" are about 200 meter thick above the brown coal.

Brown coal is used for heat production and above all for generation of electricity. About 100 million tons of brown coal are quarried yearly and are used in the beside located powerhouses.

Figure 3: Landscape of brown coal mining

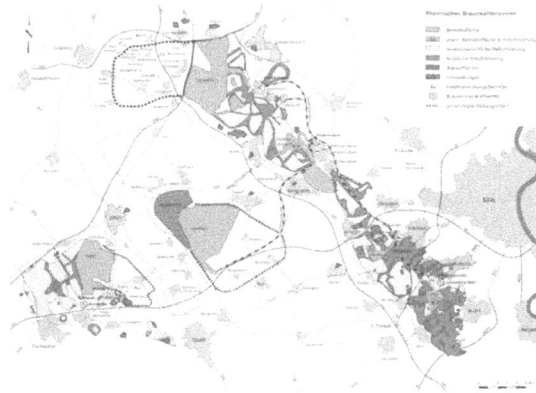

Figure 4: Area of brown coal mining

3. Introduction into the difficulty

Why are we presenting this difficulty?

We are interested into the water problems caused by the brown coal mining, because next to our country there is a location for the extraction of brown coal, what means, that there are immense consequences for our environment, population & future. In the following presentation, we are going to explain these consequences in hope to find a solution and to make this problem become popular, for the government to see that they should stop this kind of generation of energy.

4. Consequences

4.1. Consequences for the nature

4.1.1. Effects on wetlands

Wetlands, humid biotopes, homes and nature parks are exceedingly endangers induced by the removal of groundwater. Wetlands are very sensitive since there are dependent of groundwater and do not survive a too long sinking of the water. They dry out and carry of an irreparable damage.

Affected of the brown coal mining's effects is among others the nature park "Schwalm-Nette", which has international meaning.

To be found are here rare plants and animal species.

The park bestrides an area of 435 square meter additional the 789 square meter of the German-Dutch nature park "Maas - Schwalm-Nette" which is part of it since 1976. Defining component are the alder forests, which belong to Middle Europe's most endangered plant species. Those are the last big, natural grown of their kind.

Sources in this park have already dried up. Caused by the dried out surrounding area plants like alder and sedge become extinct, because they need a damp underground. Aside from that amphibian lose their spawning water. During the desiccation plants appear in the wetlands like stinging nettle, whose optimal condition then dominate. Operated by their appearance the unique wetland's type changes.

Figure 5: Sedge falling dry

4.1.2. Future of wetlands

Without retaliatory action moistness loving animals and plants become extinct and it would ensue a development towards more drought loving species. As a consequence thereof the territories' character dwindled away.

Caused by the desiccation a sinking of turf occurs and in order to that a change in the relief. Streams and ponds dry out.

Figure 6: Sinking of turf

Supporting systems could eventually preserve territories, but experts are not convinced of them yet. Those systems would have to run and function at least 80 years.

All in all the previous recirculations have impacted positive on the affected territories, but it does not succeed in all of them to establish a optimal ecological condition again. For that the territories are too heterogeneous, that means claims and requirement are different from territory to territory.

4.1.3. Follow-up processes (reuse of water)

The follow-up process of the overburden removal is the acidification of the water. The removal material consists of old residues of the then sea, like pyrite that sits in a several meter thick layer above the brown coal. When this surfaces, what is necessary when quarrying the brown coal, it oxidizes in the open air and releases acid, sulfate and iron. By precipitation and the groundwater's rise again, after ending the opencast mining, this with oxidized pyrite dosed water acidifies. In consequence of the water's transporting effect the now consisting heavy minerals stream to wells and endanger the drinking water supply of succeeding generations for hundreds of years. The residual lake, which should develop, impends to become full of acid as well.

Beside all this, those buildings become inundated, which were built beneath the originally level of groundwater, when the dewatering wells are ceased.

4.2. Consequences for water

4.2.1. Drinking water

The drinking water handed in Germany must be as specified from the requirements of the drinking water ordinance (TVO 2001), and it is regularly examined following the official example. Comparable regulations are valid in Austria and Switzerland.

Drinking water is our most important grocery but also the one that is most wasted by humanity. Farmers use harmful substances for their plants to let them grow. But the problem is that these substances remain and become an harmful environmental effect because they reduce the quality of drinking water.

An example for such a bad effect on the environment is the surface mine in Garzweiler.

It is responsible for trenching 520 m³ of groundwater, which cannot be used by humans. So it is transferred to rivers nearby. This waste has negative effects on the quality of drinking water as you can see in figure 7. The chart points out that the wasted water of the city cologne could be used by 1 milion people.

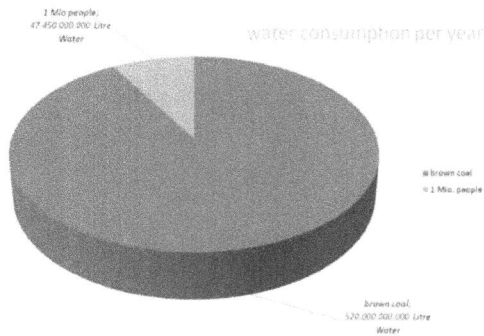

Figure 7: water consumption per year

4.2.2. Water quality

Due to brown coal mining the quality of the water is heavily reduced. One of the negative effects owing to brown coal mining is iron (sulfical) which is also known as pyrite (Katzengold), which is built at rock, that was trenched, when it has contact to the air. Through contact with air, a chemical reaction proceeds that causes an acidification of the water. It is now comparable with vinegar.

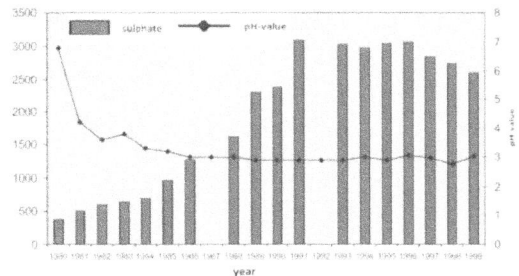

Figure 8: suflate and ph-changing

With an immense effort the water can be recycled so it can be used by people again.

The mineral and mineral water ordinance (MTVO) regulates the production, treatment and marketing of natural mineral, spring and mineral water as well as drinking water filled elsewhere. Apart from the carbon dioxide emissions from combustion of brown coal in the neighboring lignite power plants and the lowering of ground water, which leads to damage of wetlands, the open pit operation was also blamed for a number of other environmental problems, so for the acidification of the soil by the dumping of overburden and for a high particulate air pollution in the region. So also the swamp forests are threatened by the lowering of the groundwater level in the natural park Maas-Schwalm-Nette. With great effort replacement water is pumped through a system of pipes and seepage ditches in this area.

4.2.3. The Groundwater – level

Groundwater corresponds to 0.625 percent of the global waterstocks. Groundwater is defined as „underground water, which fills out the cavities from

the earth´s crust. The movement of the water is defined by the gravity and the own frictional forces. The groundwater level is exposed to an ongoing pump down/out which has a negative and also extensive effect on the environment. First new building area accrues but later it is swamped when the groundwater level rises.

Another similar consequence is mining damage which accrues though shifting an collapsing of several stratums. This occurs due to a molecular modification which is caused through water extraction of the ground.

4.3. Tectonic faults

For the extraction of brown coal in the open cast mining, ground – water has to be lowered extensively and that is the reason for tectonic faults.

Where two sediments are moving vertically against each other, there arises a tectonic fault area in the course of the years and it works as a watertight wall.

This fault is the case at the Rheindahlener crack, which runs through Rheindahlen, Hockstein and Odenkirchen. The area concerned moves herself only by a few millimeters, it nevertheless suffices to damage buildings lastingly. Besides the damage of the buildings, there are even broader difficulties, such as endangered gas pipes and the danger that sewers and other lines could break. Furthermore, there are also cracks at facades or even a lowering of an area, which is the case, when the ground – water level sinks. The Saint Magareta church in Hockstein is an example for those lowering, where the church blanket already collapsed caused by the lowering of this area and also rips can be found in the walls. For these reasons, the church hat to be restored many times, what means, that there are on one site high costs and on the other a slow destruction of a valuable building. In the end, you can say, that there are many far reaching consequences caused by the brown coal open cast mining and the sinking ground – water level.

5. Conclusion

In the end, you can say that brown coal mining is definitely a kind of generation of energy with a lot of consequences, especially for the water. This consequences regarding to water concern us and our environment. But brown coal is also an important energy source we cannot abolish in one day, so we should try to switch to alternatives like renewable energies. We hope you can understand now why this is important for us and that we could have shown you that there have to be better way of generation of energy.

References

[1] Bauer, Johann A. & Wormer, Eberhard J. : *Wasser Elixier der Gesundheit.* Lingen

[2] *Braunkohlenbericht* (2013), City of Mönchengladbach

[3] http://fossilebrennstoffe.org/fossile-brennstoffe-im-uberblick/was-ist-braunkohle/

[4] http://www.thoennessen-online.de/thoe-online/referate/braunkohle1.pdf

[5] http://fossilebrennstoffe.org/entstehung-fossiler-brennstoffe/wie-entsteht-braunkohle/

[6] http://www.hydrologie.uni-oldenburg.de/ein-bit/bilder/absenktrichter2.jpg

[7] http://www2.klett.de/sixcms/media.php/229/104103-1304.pdf

[8] http://www.bundnrw.de/themen_und_projekte/braunkohle/tagebaue_im_rheinland/tagebau_garzweiler/

[9] http://www.npsn.de/index/lang/de/artikel/1339

[10] http://www.naturparke-rheinland.lvr.de/naturparke/schwalm-nette/

[11] http://www.npsn.de/index/lang/de/artikel/33

[12] http://www.gymnasium-rheindahlen.de/

[13] http://www.netzbege.de/aktuelles/tektstoerungen.html

[14] http://www.rp-online.de/nrw/staedte/moenchengladbach/rheydt-sackt-ab-aid-1.643510

[15] http://de.wikipedia.org/wiki/Braunkohle#Umweltprobleme

[16] http://www.bund-nrw.de/themen_und_projekte/wasser/trinkwasserschutz

[17] Informations from the environment agency Mönchengladbach

Index

Social Awareness Cleans Water

Daniela Arroyave, Sebastian Ramírez, Laura Salazar, Mariana Stand

Colegio Claustro Moderno, Colombia, carolina.becerra@claustromoderno.edu.co

Abstract

Our school is located in the hillside of one of the most important mountains of Bogotá. It is supplied by a water spring rich in clean, potable water that has been provided to deprived sectors around the school. Our water status makes out of our institution a privileged spot in a polluted area. Based on this and on some studies our school has done to our water, we intend to search and implement new ways to treat water. With a special investigation in microbiology, this project will find a practical solution to prevent pollution in the source of one of the main rivers in Colombia. This spring is located in Villa Pinzón, a small town near our capital. This spring's transparent waters soon come to an end since after 360 kilometres, the water begins to look really dirty and dark. This environmental damage is caused by many tannery companies, which pose the first economic source of the region. Provided that our project gives positive results, we could propose alternative new techniques that speed up the processes of water cleaning in the river improving then the quality of potable water for all the community supplied by it.

Keywords

Immobilization, enzymes, substrate

Introduction

Being located between the Atlantic and Pacific Oceans, Colombia is a rich country in natural water resources and river systems that provide us with freshwater. Nevertheless, we also face the global weather changes and natural challenges that jeopardize such privilege. Recent droughts have been affecting different regions of our country leading to fatal consequences. But, what really concerns us is our country's weakness in water treatment and preservation policies. Despite all our resources, there are still sectors that are not able to obtain freshwater in their homes. It should be a granted right, but unfortunately for many families, it remains a dream. This is due to many factors, poor water technologies and water pollution as a result of uncontrolled industry waste regulations are probably the strongest. This problem will not change overnight, but it is in our hands to find viable solutions that can be used in our environment. This is why we chose to work on the water problem in Villa Pinzón. It affects us directly for the water of this river is one of our capital's main source of water. We expect to have positive results in our research as well as to raise social awareness in our community so that many of us can start preparing towards the future, innovating and learning from other countries that have used cutting-edge technologies to overcome such challenge.

1 Objective

There is a technology that uses microorganisms grown in polymers to generate enzymes that are able to affect the main structure of the dyes causing the cleansing of the water. We want to follow an observation process.

Based on the former information, we took an experiment made in order to decolorize and have a detoxification process in water with textile dyes and decided to apply it to polluted water taken from the water source in Villa Pinzón. The purpose of our experiment is to confirm whether the positive results obtained in water with dye residuals apply as effectively to water polluted with chrome residuals from the tanneries surrounding Villa Pinzón.

2 Methodology

Microorganisms such as *Phanerochaete chrysosporium* are able to help in the processes of oxidation of certain organic pollutants. This type of white rot fungi seems to have the power to break chemical chains that pollute water. It is also key in carbon cycles. They can decompose explosive contaminants and toxic residuals. We need then to grow this fungi in order to put it in contact to the polluted water and see if it works in the same way with chrome as it does with textile dyes. The procedure we will follow should be the same.

We are attempting to work along with a university laboratory in order to have access to all their resources so that the quality of the experiment is not affected.

2.1 Materials

4 Petri boxes with bran extract (agar) for the fungi growth;1 250 ml Erlenmeyer flask with 50 ml of bran extract and 20 1cmx1cmx1cm polymer cubes; 1 Pasteur pipette; 1 big funnel ; filter paper, tweezers; 1 250ml Erlenmeyer flask with 100ml sterile distilled water; 1 250 ml glass beaker; 1 13x100 test tube with 0.1 of residual water; ; 1 13x100 test tube with 0.1 of distilled water (Bradford); ; 1 13x100 test tube with 12 Bradford reactor.

2.2 Fungi growth

The fungi needs to be grown for four days. You should spread it in agar formula at 30° C.

2.3 Immobilization process

Set a 250 ml. Erlenmeyer with 50 ml bran extract and 15 polymer cubes. Add 3 boxes of grown fungi. Leave in incubation for 8 days at 30° C at 150 rpm.

2.4 Fermentation

Add the residual water. Filter the media with the immobilized fungi. Wash the cubes with the distilled water to clean the cells excess. Observe immobilization. With the tweezers and next to the burner, pass 5 cubes of immobilized fungi to each Erlenmeyer flask. Filter the extract and determine units of final color and proteins. Take to incubation for 8 days at 30° C at 150 rpm in the field with the immobilized fungi.

3 Results

This is the furthest we have reached in the experiment. We are waiting for the process to finish so that we can give the conclusions.

In case it would not work, we would like to try algae found in the waters of our school since they have provided a special environment for certain species to keep alive.

Acknowledgements

We would like to thank Nazly Villamil and Diana Gaitán for their support and guidance through this process. They have helped with their knowledge and experience in the field.

References

[1] Applied and Environmental Microbiology. (2000, Aug). *Decolorization and Detoxification of Textile Dyes with a Laccase from Trametes hirsute*, p.3357-3362

[2] Fungi Immobilization on Synthetic Base by Absorption guidelines. (2005, Apr).

[3] National Center for Biotechnology Information, U.S. National Library of Medicine. (2008, Dec). *The white-rot fungus Phanerochaete chrysosporium: conditions for the production of lignin-degrading enzymes.* Retrieved from *://www.ncbi.nlm.nih.gov/pubmed http /18810426*

[4] Maya Achicanoy Diana Marcela, Universidad de los Andes, (2004). *Estudio de Alternativas de Desinfección para el Control de Patógenos en el Río Bogotá.* Retrieved from*file:///C:/Users/Carolina/Downloads/Alternativas desinfecci%C3%B3n%20_r%C3%ADoBogot%C3%A1 .pdf*

ECOSYSTEMS: BIODIVERSITY OF THREE LAKES IN PIEDMONT

Andrea Lamberti, Elisa Lombardi, Simone Pelizzola, Sofia Zompi

Liceo Scientifico "Carlo Cattaneo", Turin, Italy, g.pace@libero.it

Abstract

We live in Piedmont, a region in the north-west of Italy, this area is characterised by the Alps and Pianura Padana where the longest Italian River, the Po, flows with its tributaries.

For these reasons the area is quite rich of water thanks to the glaciers and the snow on the Alps, moreover the orography of our region well suited the presence of water to be used by humans.

We are very concerned to protect the ecosystem and biodiversity in particular.

Our research aims to analyse the water ecosystem of two lakes present in our region, Avigliana and Viverone lakes.

The research will take into consideration the flora and fauna, the physic/chemical features of the water and the geological history of the area they are in. The wide range of living forms inside the lakes and along their banks show how important water is for life.

Keywords

biodiversity, saprobic, ecosystem

The purpose of our investigation

The purpose of our project is to find out if the quality of the lake water is suitable to maintain the biodiversity and/or which part of it is in danger. We divided our research in four different phases

At first, we took our data about the surface of Avigliana and Viverone lakes. In this phase we selected these data and we made a presentation with them. They refer to the two lakes' fauna and flora, so we analysed the typical plants which live in those areas and the animals which made up, with them, Avigliana and Viverone's environment. After that we looked for plankton and microorganisms living in those lakes aiming to understand the environmental situation of both lakes. In fact this analysis let us understand if the lakes are clean or not, because there are some particular species of bacteria which proliferate in clean water.

Then we went to Candia Lake to take data in a different period of the year. We took some water aiming to see the bacterium species which live in it and to understand the lake environmental situations.

Besides we measured the lake depth and the depth at which the light is reflexed and then refracted with an angle that doesn't allow the light to return to the top. We used a particular instrument which is called Secchi Disc. It has alternates black and white strips so that all the light is reflexed by it or all of it is captured but it's not divided. We put it into the water and when we couldn't see it any more we measured the depth. So we knew the depth of the "Photic Zone", where there is no light. We did it because if this depth is too little it means that the lake is dirty. From those measures and the analysis of the bacteria made with an optical microscope we understood that Candia Lake is one of the cleanest of Piedmont. The same day we went to analyse the flora of the lake and we took some plants to make a herbarium.

The third phase consisted in producing our presentation in which we put the analysis of Avigliana and Viverone lakes. We engaged ourselves to find geographical and historical information about those lakes and to make them easily readable by everyone.

Finally the fourth phase will be the journey to Singapore where you'll hear from us everything we discovered!!

METHOD OF INVESTIGATION

Table 1

SAPROBIC INDEXES	POLLUTION LEVEL	SAPROBIC CLASS
1,0-1,5	None. Swimming Water	I
1,6-2,5	Moderate. Swimming Water	II
2,6-3,5	Dangerous. Not Swimming Water	III
3,6-4,0	Very Dangerous. Not Swimming Water	IV

To calculate the level of pollution in the lake we looked for some particular species of bacteria which live in it.

Then we divided it in bacteria of 1st, 2nd, 3rd, 4th class.

In fact some bacteria proliferate in waters with specific clean levels.

After counting how much species of each class we had found we used this data to calculate the Saprobic Class of the lake with this algorithm.

$$IS = \frac{4D + 3C + 2D + A}{D + C + B + A} \quad (1)$$

where D is the number of bacteria of class IV, C is the number of bacteria of class III, B is the number of bacteria of class II and A is the number of bacteria of class I. IS is the Saprobic level.

With all the results of our experiments we made an average and we found the result

$$IS=1,77$$

It means, looking back to Table 1, that the lake has Swimming Water and it Pollution Level is not elevated and Candia Lake is one the cleanest of Piedmont and one of the most clean in Italy.

FIGURES

Here's some photos taken when we went to Candia Lake.

Figure 1 Some of the instrument used for our research e.g. Secchi Disc (right).

We also saw some microorganisms using microscopes and we took photo of them directly from the microscopes.

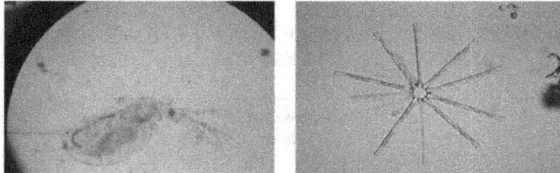

Figure 2 Microoragnisms in Candia Lake.

We took photo of plant life. We observed some typical fauna when we were withdrawing some water to examine the level of pollution of the lake.

Figure 3 Flora and fuana in Candia Lake

Conclusion

We can surely say that the lake of Candia is one of the less polluted lakes in northern Italy. We can understand it from the level of saprobic microorganisms which are present in the lake and with our investigation on it, observing them with the microscope. To keep the environment pure we must avoid introducing foreign specie that can destroy the food chain and the environment also. For example recently in the lake are being found some prawns that had been introduced in the lake and these animals are destroying the water plants which live in the lake and these prawns are also difficult to eat by the predator for their aggressiveness.

The work of the guard of the lake is very important to keep the lake clean and liveable but every one of us has the duty to not pollute it.

Acknowledgements

Ente provinciale del Lago di Candia

Professoressa Rita Cavallone

Dirigente scolastico Professor Sabatino D'Alessandro

References

[1] F.Bona, A.Maffiotti- L'eutrofizzazione del lago di Viverone: studi e proposte d'intervento- Regione Piemonte

[2] Situazione ecologica e proposte per il risanamento dei laghi di Avigliana e della palude dei Mareschi- Regione Piemonte e Istituto di zoologia Università di Torino.

Research of Wuhan Lakes' Anastomosing and Urban Water Ecosystem

Yuke Zhou, Yixin Xu, Yuxin Chen, Jorryn Wu

Wuhan Experimental Foreign Languages School, China, 851844208@qq.com

Abstract

Earth is the only planet known to have abundant liquid water on its surface among all of the planets in our solar system. Water covers about 71% of Earth's surface. This kind of water is called the global ocean. The global ocean is part of Earth's Hydrosphere. The hydrosphere is the portion of Earth that is water. The three other major spheres of Earth are the biosphere, atmosphere and geosphere. The biosphere is the part of Earth where life exists. For billions of years, livings are maintaining excellent biological balance. But during Mankind's short life, livings' biological balance is crashed rapidly. The normal life of mankind relates to the whole ecosystem gravely, and it is influenced greatly by this crashing. This paper takes the linking of urban rivers and lakes as the research object. It shows the process below: observing the ecosystem of Wuhan river areas, building lakes' anastomosing model, making hypothesis of ecological problems, testing this hypothesis, analyzing results, communicating results and drawing the conclusions. This research aims to find the feasible scheme of anastomosing urban rivers.

Keywords

Ecosystem, Anastomosing Rivers

1 Introduction

Known as the city of rivers, Wuhan is located in mid-Hubei Province. With many years of urban development, the problem of lake pollution is becoming more and more serious, which threatens the safety of residents' drinking water. With China's longest river - the Yangtze River flowing through the city, Wuhan is now carrying out the project of "Lakes' Anastomosing", the Yangtze River and the pollution of the lake is about to flow communicating, forming a large network of aquatic ecosystems in order to reduce pollution index, effective solution the purpose of lake pollution problems. This study was designed to test water quality analysis, to understand and summarize the six lakes connected Wuhan impact on urban water systems, rationalization proposals put forward to improve the project, in order to effectively alleviate the ecosystem pollution problems.

2 Method of the investigation

- Building the lakes connectivity aquatic network model.
- Gathering the water samples of Yangtze River and lakes.
- Injecting the water samples into the model's channels.
- Imitating lakes' anastomosing.
- Testing the changing of water quality by means of chemical reactions.
- Comparing the DTP contents of the lakes and rivers' samples.

3 Results of the experiment

DATE	SITE	DEPTH	TN (mg/L)	NO3-N (mg/L)	NO2-N (mg/L)
Mar.17th 2014	Lake	0.500	2.484	1.349	0.167
Mar.17th 2014	Anastomosed lake	0.500	1.347	1.030	0.020

Table 1 The Content of TN, NO3-N and NO2-N in Water Samples

Eq.(1)DTP, mg/l=m*100/v

Eq.(2)DP, mg/l=m*1000/v

According to experimental data, the water in anastomosed lake is cleaner than that in the original lake, which proves the effectiveness of lake anastomosing.

It also shows the necessity of building water treatment plants downstream, which will help to improve the purification of water in lakes.

References

[1] NATIONAL URBAN WATER CONSERVANCY
ACADEMIC SEMINAR THESIS ALBUM [2003]

The value of wetlands as a retarding basin and as an inclusion in urban areas

Brendon Aung, Calvin Fletcher, Abigail Hyndman, Heshala Wijerathna

John Monash Science School, Australia; adriana.abels@jmss.vic.edu.au

Abstract

Wetlands are extremely important ecosystems which provide a wide variety of uses for both people and wildlife. Wetlands improve water quality by filtering the water. Its natural filtration process removes excess nutrients before the water flows from the wetland into nearby rivers and creeks, bays and oceans.

Wetlands are a significant economic and social benefit as they help in flood control. They act like sponges and absorb excessive rainfall. They provide a habitat for many animal species; particularly providing a refuge for threatened species such as frogs and migratory birds. Recently it has also been discovered that wetland areas are excellent carbon sinks.

In our project we would like to explore the policy of wetland inclusion in open spaces or in urban areas and therefore the benefits associated with wetlands in the long term. Our aim is to create a wetland (or water storage area) at the corner of our school site where, during heavy rainfall, substantial volumes of water accumulate. This wetland will provide a pleasant environment for the students of our school, a habitat for local wildlife; but most importantly, an area which contributes to increasing wetland spaces in urban areas where the water can be filtered before it eventually enters the nearby waterways.

Keywords

Wetlands, CO_2 sinks, ecosystems, biodiversity, sustainability

Introduction

Wetlands are areas of land that are saturated by fresh/salt water either all year or during a certain time of the year (changing seasons). Such natural features are important in sustaining our ecosystem as wetlands support a wide range of plants and animals including humans (two thirds of the fish on the planet come from wetlands). Interestingly, the amount of water in the soil also determines the amount of animals and plants living in wetlands, including aquatic and terrestrial species. As a result, such an environment can promote the growth of unique plants and animals as well as maintain the development of lush wetland soils. Wetlands can be found all over the world regardless of the climate as demonstrated by the spread of wetlands from tropical locations to Polar Regions. Each type of wetland can support their own unique set of organisms.

How do Wetlands form?

Most natural wetlands are formed when excess rain or water from watersheds cause flooding. This excess water combined with poor drainage, and low lying land contribute to the formation of wetlands. Many wetlands were formed during the ending phase of the last ice age. This was when large glaciers melted and low lying areas and depressions quickly filled with water. Sediment and other organic substances then quickly filled the depressions with water giving rise to adapted soils and environments to suit a wide variety of unique plants and animals.

Features of Wetlands

For any water body to be defined as a wetland, there are a few unique features the water body must have. To put it simply, a wetland is land that is immersed in water for a long period of time such that it supports aquatic plants that are specially adapted to wet conditions and produces hydric soil (soil that has been saturated with water for so long it becomes anaerobic). Thus, many other similar water bodies such as watersheds can be distinguished from wetlands.

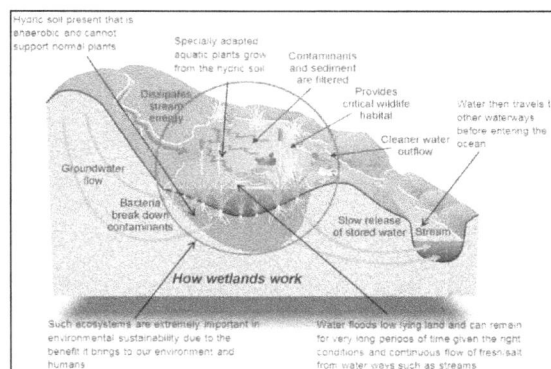

Figure 1: The functions of wetlands and some critical features.
Source: http://nmfarmgirl.umwblogs.org

There are various types of wetlands and these can be summarized below.

Figure 2: Differences and similarities between the various types of Wetland

Wetland type	Major Subtypes	Soil type	Biodiversity	Locations	Water Frequency	Water Type
Marshes	Tidal (coastal) marshes Non-tidal marshes	Mineral soil	Largest due to balanced pH and nutrients Supports herbaceous (non-woody) plants	Mouth of rivers, coastal places (Tidal marshes), Boundaries of other water bodies (Non-tidal)	Continuously filled or frequently filled with water	Salt or Fresh from surface water
Swamps	Forested Swamps Shrub Swamps	Mineral Soil	Large Biodiversity Supports woody plants, trees and shrubs	Low-lying regions next to rivers and streams.	Certain times of the year only	Salt or Fresh
Fens	N/A	Organic soil (peat)	Larger biodiversity compared to bogs Grasslike plants & shrubs	Can be found next to bogs More often in northern hemisphere and associated with low temperatures		Water from ground and surface water (Neutral or alkaline)
Bogs	N/A	Organic Soil (peat)	Lowest biodiversity due to poor soil nutrition Trees, shrub, moss, carnivorous pitcher plant	Can be found next to fens More often found in northern hemisphere and cold temperatures		Fresh water from Precipitation (acidic)

Benefits of Wetlands

Economic benefits

The presence of wetlands connected to rivers and streams can significantly improve the water quality. Suspended sediments and nutrients in the water settle to the wetland floor and are absorbed by plant roots and microorganisms as the water current slows. Consequentially algal blooms are prevented as they lack the required nutrients. Such growths can reduce fish numbers and leave the water unfit for use. The wetlands can replace large water treatment plants as an alternative for purifying water.

In regions prone to flooding, wetlands can act as water sinks due to their large inlet and small outlet. Excess water during storm and heavy rain events begins to return to the atmosphere as a result of evaporation or plant transpiration. To put it simply large amounts of water are released slowly downstream rather than quickly preventing destruction of land. Alternatively wetlands can act as buffer zones reducing risks of bushfires which has become a greater issue for Australians in recent years. In urban environments, wetlands can cause a cooling effect during warmer months while moderating stronger winds as they come about.

Most importantly wetlands can act as carbon sinks where they regulate many exchanges with greenhouse gases. Carbon is stored through photosynthesis leading to an accumulation of organic matter in plant biomass and soil. Wetland plants tend to grow faster than they decompose leading to a net reduction of CO_2 levels. Although they do sequester carbon, evidence has shown in wetter climates other greenhouse gases such as methane and nitrous oxide are produced as a result of anaerobic conditions, though wetter climates also tend to lead to larger scale carbon sequestration.

Environmental benefits

Freshwater wetlands hold more than 40% of the world's species, 12% of which are animal species. Particularly reliant are frogs and other amphibians. Amphibians are in a state of decline due to increasing arid conditions and loss of habitat. They must be near a source of water to survive as desiccation will lead to death. Likewise for reproduction amphibians need a source of slow flowing water where they can lay their gelatinous eggs with minimal disruption. Wetlands also serve as nurseries for the larval forms of these amphibians where resources and food are abundant.

Similarly ducks, geese and other water birds heavily rely on wetlands for nesting, breeding and moulting. It is vital during the drought period of the summer that wetland habitats are available for water birds to retreat to as other areas begin to dry out. The reduction of water flow speed due to wetlands also makes nearby streams and rivers more favourable as habitats for these birds.

Social benefits

Wetlands also fulfil a recreational purpose. The addition of wetland habitats in local parks can add a new aesthetic while also cooling the local environment. Improving these amenities in such a way leads to increased popularity and thus encourages more people to leave their houses. In urban areas these wetlands can allow for people to interact further with nature. Furthermore the wetlands can inform and educate the community about wetland ecology and the importance of wetlands both economically and environmentally.

Bird watching is one major recreation activity associated with wetlands. Bird watching allows people to get up close and personal with wildlife while also providing exercise and reducing stress. Alzheimer's Australia state that activities such as bird watching are valuable in order to keep the mind and body active. In Victoria wetlands house a myriad of different avian varieties from reed dwellers such as the Dusky Moorhen, to shorebirds including the Curlew Sandpiper. Through bird watching members of the local community can increase their understanding of issues surrounding water dwelling avian species such as their general need for a wetland habitat. Bird watching can also be beneficial for data collection as demonstrated by the initiative 'Climatewatch' where citizens can take a photo of a particular species and log where it was spotted on a smartphone application. This in turn can be useful for scientists to determine the long term effects of our activity on certain species of bird.

Furthermore the amalgamation of people to form a 'friends association' where regular citizens involve themselves with the maintenance and care of the wetland can result from the presence of a wetland in an urban area. Support for such a group will increase community engagement while giving leadership opportunities for those who wish to accept them. A friends association can raise awareness for the benefits of wetlands amassing interest in the subject among others. Awareness is very important as funding for wetland restoration and research is essential for the future of worldwide wetland communities.

Wetland Policies

It is essential that wetlands are protected not only by humans, but by law. The Australian Wetland Policy is based on the Ramsar Convention, Iran, 1971. Ramsar is an Iranian city, which hosted the "Convention on Wetlands of International Importance". The policy strives to promote

and protect wetlands all over the world. The convention became ratified in 1975.

The Australian policy based on the Ramsar Convention is divided into six articles. These six articles cover:

1. Basic information on measures taken by contracting parties
2. Further information on wetlands designated for the list of wetlands of international importance
3. Wise use of wetlands
4. International cooperation
5. Wetland reserves and training
6. General comments on the convention and its implementations

This policy was originally designed in 1971, to protect waterbird habitats. Australia was one of the first countries to become a signatory in the same year, even before Ramsar broadened its policy to cover wetlands.

Although many would believe that the sole purpose of the Australian Wetlands Policy is to protect wetlands and the environments surrounding, the reality is that there is policy to protect and policy to remove wetland environments from the "List of Wetlands of International Importance". However, so far, no Australian wetland has ever been deleted from the list. The objective of this list is to *"develop and maintain an international network of wetlands which are important for the conservation of global biological diversity and for sustaining human life through the maintenance of their ecosystem components, processes and benefits/services".*

Internationally, the list protects 2,122 sites (known as Ramsar Sites). Combined, these sites cover approximately 205,366,160 hectares. This is an increase from the 1,021 Ramsar sites that were listed in the year 2000. Over the space of 13 years, more than one thousand wetlands have been recognised for their significance internationally.

There are many International and National organisations protecting Australian Wetlands such as Ramsar and Wetlands International. Wetland Protection Organisations that work with the Commonwealth government include the Department of Sustainability, Environment, Water, Populations and Communities (SEWPaC), CSIRO Land and Water, Murray Darling Basin Authority and Land and Water Australia.

There are also many Non-government organisations, such as: Bird Observers Club of Australia, Conservation Volunteers, Australian Wetlands Alliance, Nature Conservation Council, Wetlands Environmental Education Centre, Wetland Care Australia, Wetlands International, The Wilderness Society and WWF Australia.

A Wetland in an Urban Environment – The Royal Park Wetlands

Trin Warren Tam-boore (TWTB) is a created wetland in the north western corner of Royal Park (Melbourne, Australia). It was designed to treat and recycle stormwater which had run off from the roads, rooftops and gutters.

TWTB was created in 2005, is 170 hectares and consists of many parts which all act to treat the stormwater. The TWTB is home to many flora and fauna and is only a few minutes from Melbourne.

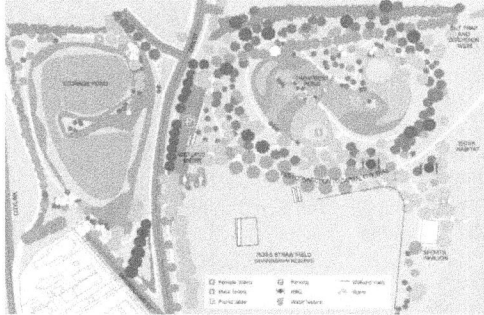

Figure 3: a bird's eye view of the Royal Park Wetlands

Source: www.thatsmelbourne.com.au.

The wetland consists of a self-guided walk which teaches the general public about urban wetlands and how each stage of treatment happens. TWTB has a treatment pond (Figure 3: number 3, 4 & 5) in the shape of an 'S' and a storage pond, where the treated water is stored until usage. The 'S' shape of the treatment pond and its changing depths ensure that waters passes through at a slow rate. This way, gaining the maximum exposure to the cleansing effects of plants and sunlight. When the water is in depths less than 0.5m, the plants and associated micro-organisms absorb nitrogen and other unwanted nutrients in the water.

Once the water has gone through the treatment pond it flows through underground pipes until it reaches the storage pond (Figure 3: Number 7). This water is then used for irrigation as Class A water, or piped to other locations for watering. Further to treatment of water the TWTB, this area also provides a great habitat for natural flora and fauna. Home to 270 bird species, lizards, fish, and insects, the wetland has created its own ecosystem. This wetland provides food, water, shelter and breeding grounds to a wide range of organisms. According to ecologists, the TWTB has been a great success and they are hoping and encouraging more areas like this are set up around urban Melbourne.

CONTENT

The purpose of the investigation

Wetlands are a key part of our environment and can be found all over the world. There are many different types of wetlands with a commonality being that they provide a primary habitat for thousands of species of birds, fish, amphibians and insects including species unique to wetlands. Alternatively, wetlands provide water quality improvements, are essential for the detention of floods, provide a water filtration system and provide a haven in the suburbs.

Method of investigation

After completing substantial research on the environmental, economic and social benefits of wetlands, we decided to construct a wetland on our school site. The purpose was to not only boost environmental sustainability by providing a habitat for urban birds and frogs; but also to provide an oasis for the students in our school: a passive recreation space.

Although the site was chosen in consultation with Monash University Engineering Department and Monash Garden Maintenance, we had discovered during our research that the site we were thinking of converting into an urban wetland had originally been a swamp. It was the lowest point of the Monash University Campus and consequently, the majority of rain water would accumulate at this site. In fact, several large drains had been constructed in the area to cater for the storm water runoff. As such, this location proved to be the ideal spot for our wetlands.

Figure 4: rain water accumulating in the dish drain which leads to our wetland.

The accumulation of water in the dish drain (Figure 4) highlights the practicality of our wetland location. As the gutter becomes filled by more and more water, the excess water will then run off the side and travel down the hill to our wetland which can be seen at the top right-hand corner. Furthermore, as highlighted by the image below (Figure 5), the trees present are able to block out the main portion of the wetland from direct sunlight. As such, this prevents rapid evaporation of the water and allows water levels to be maintained for longer periods.

Figure 5: View of the wetland with shade from existing trees.

Our next task was to assess the movement of water through the area and decide the best place for the 'pond'. We were fortunate that we had some rainfall one weekend and in collaboration with Monash, we were able to commence construction of our wetland. The shape of the pond and rivulet were painted onto the ground and earth moving equipment was dispatched to the site to commence the digging process. Within the hour our pond, of about 40 centimetres deep and a diversion channel had been dug.

Figure 6: The pond at the end of the channel which is damp from the downpour on Sunday March 16 2014.

Results of the experiment

We were not at school on the weekend of rain, and upon our return on Monday, we noticed the damp patch (Figure 6). Although the weather bureau has predicted rain, it has not been forthcoming. So, in order to stabilise the channel and the pond we decided to use a hose to temporarily fill our wetland. Figure 7 portrays roughly what our wetland should look like. Water would be moving from the gutter down the channel naturally until it reaches the main 'hole' where it would fan out into the space.

Figure 7: Water flowing down the channel to the pond at the end.

The strategic shape and depth of our wetland allows water to seamlessly run along the rivulet to finally arrive at the main 'hole' (the circular damp patch). This is due to the natural elevation of the ground allowing the water to move down to lower elevation.

We commenced researching our project in early February, and the wetland was dug in early March; a very short amount of time to gauge its effectiveness and right in the middle of a hot and dry Melbourne summer; however, we feel that we have achieved some success in selecting the most appropriate site for our wetland.

Figure 8: The dish drain with rain water and oil pollution.

Figure 8 highlights the pollution present in the water. As can be seen in Figure 8, we received some rain recently and it was evident that the water built up in the car park was carrying unwanted grease and oil left from vehicles. The wetland would then serve to filter the water from pollutants which would benefit the flora planted around the wetland.

We are currently in the process of sourcing indigenous plants to the region: those that are water tolerant and those that are appropriate for the levee banks. We see this as an ongoing project and we are looking forward to continue assessing the effectiveness of our wetland, both in filtering water and promoting biodiversity. We are hopeful that these results should become evident with the winter rains.

Acknowledgements

Ms Abels	Facilitator – Head of Humanities
Ms Chaplin	Facilitator
Mr Chisholm	Assistant Principal
George Buchanan	JMSS Handyman
Mark Corea	Monash University
Diggers	Workers provided by Monash University
Mrs Corkill	Consultant and plant advisor

Bibliography

[1] Alzheimers Australia RSS. (n.d.). *Bird Watching in My Back Yard*. Retrieved February 25, 2014, from <http://www.fightdementia.org.au/get-involved/bird-watching-in-my-back-yard-ebook.aspx>

[2] Author Unknown (1999). The assessment of restoration of habitat in urban wetlands. *Landscape and Urban Planning, 43*(5), 227-236.

[3] Birds in Backyards. (n.d.). *Benefits of Bird Watching for Your Family*. Retrieved February 26, 2014, from <http://www.birdsinbackyards.net/Benefits-Bird-Watching-Your-Family>

[4] Briney, A. (2014). *Wetlands*. Retrieved February 26, 2014, from <http://geography.about.com/od/physicalgeography/a/wetlands.htm>

[5] City of Melbourne. (n.d.). *Trin Warren Tam-boore*. Retrieved February 9, 2014, from <https://www.melbourne.vic.gov.au/AboutMelbourne/ProjectsandInitiatives/MajorProjects/Pages/TrinWarren.aspx>

[6] Chovanec, A. (1994). *Man-made Wetlands In Urban Recreational Areas - A Habitat For Endangered Species?*. *Landscape and Urban Planning, 29*(1), 43-54.

[7] Department of the Environment. (2012). *The Role of Wetlands in the Carbon Cycle*. Retrieved February 7, 2014, from <http://www.environment.gov.au/system/files/resources/b55b1fe4-7d09-47af-96c4-6cbb5f106d4f/files/wetlands-role-carbon-cycle.pdf>

[8] Department of the Environment. (2013). *Planning and management of urban and peri-urban wetlands in Australia - Fact sheet*. Retrieved February 8, 2014, from <http://www.environment.gov.au/resource/planning-and-management-urban-and-peri-urban-wetlands-australia-fact-sheet>

[9] Department of Primary Industries (n.d.). *Wetland habitats*. Retrieved February 7, 2014, from <http://www.dpi.nsw.gov.au/fisheries/habitat/aquatic-habitats/wetland>

[10] Devitt, T. (n.d.). *Wetlands and floods*. Retrieved February 7, 2014, from <http://whyfiles.org/107flood/4.html>

[11] Discovery Communications. (2011). *What causes wetlands to form?*. Retrieved February 26, 2014, from <http://curiosity.discovery.com/question/what-causes-wetlands-form>

[12] Ehrenfeld, J. (2000). *Evaluating Wetlands Within An Urban Context. Ecological Engineering, 15*(3-4), 253-265.

[13] Koch, F., Andrews, R., and Murray, C. (n.d.). *Hydric Soils*. Retrieved February 26, 2014, from <http://courses.soil.ncsu.edu/ssc570/student_projects/571_web_page/hydric.htm>

[14] Melbourne Water. (n.d.). *Constructed wetlands guidelines*. Retrieved February 9, 2014, from <http://www.melbournewater.com.au/>

[15] Parks Victoria. (n.d.). *Wetland birds*. Retrieved February 25, 2014, from <http://www.enviroactive.com.au/wetlands/birds>

[16] Ronca, D. (2008). *How Wetlands Work.* Retrieved February 13, 2014, from <http://science.howstuffworks.com/environmental/green-science/wetland2.htm>

[17] That's Melbourne. (n.d.). *Royal Park Wetland.* Retrieved February 9, 2014, from <http://www.thatsmelbourne.com.au/Documents/Parks/RoyalParkWetland.pdf>

[18] United States Environmental Protection Agency. (2006). *Economic benefits of wetlands.* Retrieved February 7, 2014, from <http://water.epa.gov/type/wetlands/outreach/upooad/EconomicBenefits.pdf>

[19] United States Environmental Protection Agency. (2012). *Fens.* Retrieved February 13, 2014, from <http://water.epa.gov/type/wetlands/fen.cfm>

[20] United States Environmental Protection Agency. (2001). *Types of Wetlands.* Retrieved February 13, 2014, from <http://water.epa.gov/type/wetlands/outreach/upload/types_pr.pdf>

[21] Wetland Care Australia. (n.d.). *Localised Benefits of Wetlands.* Retrieved February 8,2014, from <http://www.wetlandcare.com.au/docs/education/IB%20Localised%20Benefits%20of%20Wetlands.pdf>

[22] Wetlands International. (2013). *What are wetlands?.* Retrieved February 13, 2014, from <http://www.wetlands.org/Whatarewetlands/tabid/202/AlbumID/11392-86/Default.aspx>

[23] WWF. (n.d.). *Freshwater wetlands.* Retrieved February 7, 2014, from <http://wwf.panda.org/about_our_earth/ecoregions/about/habitat_types/habitats/freshwater_wetlands/>

[24] WWF. (n.d.). *The value of wetlands.* Retrieved February 8, 2014, from <http://wwf.panda.org/about_our_earth/about_freshwater/intro/value/>

[25] WWF. (n.d.). *Types of wetlands.* Retrieved February 13, 2014, from <http://wwf.panda.org/about_our_earth/about_freshwater/intro/types/>

Wastewater Management:

Algae for Wastewater Treatment and Water Education

Shengzhi Wang, Gabrielle Kamie

Makuhari Senior High School, Shibuya Kyoiku Gakuen, Japan: kokusai@shibumaku.jp

Abstract

This purpose of this project is about wastewater management solutions. By providing solutions from scientific and social aspects, our project allows a wide array of people to participate, producing effective ways to solve wastewater problems in our societies.

In the scientific area, we will explore the potential of algae, and how it will contribute significantly as an eco-friendly way to treat wastewater, while at the same time producing a clean energy source — biofuel.

Since domestic water usage is one of the main causes of wastewater, increasing education on water usage is essential. The social area focuses on how education can raise awareness on the ever-increasing wastewater problems.

Keywords
Wastewater Management, Water Education, Algae, Energy

1   Introduction

As earth's population increases as it has never done before, providing wastewater management solutions and sustainable energy is more important than ever. Our project aims to discover wastewater management solutions by finding ways to utilize water resources to treat wastewater and provide clean energy. Our other goal is to emphasize the importance of better water education that encourages smarter water usage.

2   Alga's Use to Improve Wastewater Management

2.1   Potential

The replacement of fossil fuels is one of the biggest issues we face today as the earth's environment and climate deteriorates. Bio-fuel has been shown as one of the most promising alternatives as a sustainable fuel that can be produced with earth's existing resources. Bio-fuel is produced from organic substances such as corn, soybeans, and algae. However, most of the resources needed to create the fuel is agricultural products we rely on every day, producing bio-fuel could be the cause of rising food prices. That concern is non-existent with the use of algae [3]. Algae can be used without the fear of interfering with food supplies [1]. In addition, it has been shown that algae and micro-algae can contribute significantly as a way to clean wastewater, while using the nutrients in it foster growth.

2.2   Method

Algae will be produced using energy from sunlight, carbon

Figure 1: NASA OMEGA project's Photobioreactor

dioxide in the air, and nutrients in the wastewater. It will be done on land based facilities (see fig. 1) or offshore reactors. The result of this process will be clean water and algae. The cleaned water will be brought back to rivers or the ocean and the resulting algae will be used as the raw material for producing bio-fuel. The algae will be made into crude oil which will then become fuel suitable for cars or planes [2].

2.3   Effectiveness

Alga is highly efficient in terms of area usage. In terms of land/water area usage, micro-algae have the potential to be roughly 1,000% to 10,000% more efficient than other biomasses (see fig. 2). In terms of cleaning wastewater, it uses relatively little area on the water to process large

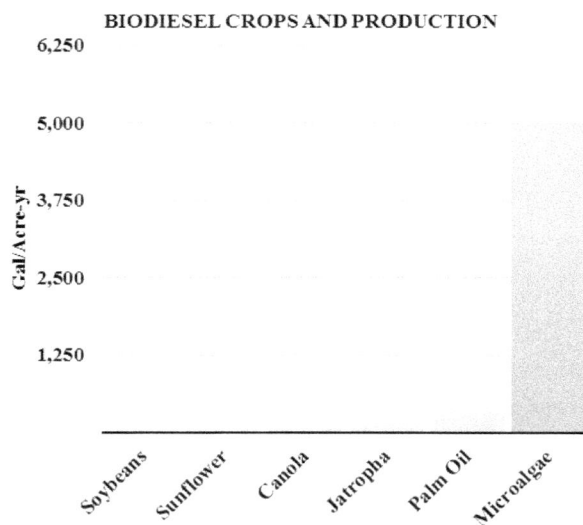

Figure 2: Micro-algae biofuel production's Gal/acre-yr compared to those from other biomasses [1]

amounts of wastewater. For example, San Francisco's daily 65 million gallon (or roughly 250,000 kL) of wastewater can be processed by just 1280 acres (about 5.2 km²) of algae facilities, according to NASA scientist Dr. Trent's research [1].

2.4   Result and Discussion

The production of algae can serve a vital role in cleaning wastewater without using harmful chemicals. In addition, producing and using its biofuel will not increase the carbon dioxide in the atmosphere and thus will maintain carbon balance. However, the process is relatively new and has made little headway except in cases of scientific research, and requires significant investment to produce such facilities. Increased investments maybe expected once the costs are lowered to more reasonable amounts.

3   Social Solutions

3.1 Raising awareness on wastewater is important

Domestic usage is one main reason for the amount of wastewater produced everyday. Therefore, educating children and raising awareness on the current situation of wastewater is one of the most effective solutions. Acting socially requires the help of citizens, and this can only be effective if a vast number of people understand the potential behind it and participates. Although advanced technologies have proven to be making a big impact on wastewater management, this project aims to give the majority of the public ideas on how to participate as well.

3.2   Teaching Methods

We plan to hold lessons for students and raise wastewater awareness. Throughout these lessons, we will inform them about the current situation of wastewater management, as well as the kind of cooperation that is needed of people to secure our access to safe water in the long term. This includes research regarding data of how much wastewater is produced every day and how it will be affecting upcoming generations of water users. We will also introduce ideas on how to conserve water through activities such as quizzes, inspired by The Clean Water Education Partnership [5], about replacing everyday-used items with eco-friendly, less harmful items. Finally, to raise awareness, we will design concept posters to put around the school. These lessons will be held annually to elementary school students.

3.3   Result

Students who take the lessons will understand the importance of water for human beings. They will be more likely to incorporate water-conserving actions into their lifestyle, making more eco-friendly decisions. They will learn to provide the next generation with the same amount of clean and healthy water resources.

4   Conclusion

By providing a way we can combat the increasing amounts of wastewater around the globe, as well as providing a way to decease it on the user's end, we have made a persuasive argument on the topic of wastewater. This project will successfully promote water conservation and spark interest in the future of algal water treatment methods.

Acknowledgements

The authors would like to thank Mr. Ueda and Mr. Takahashi from the 「一般社団法人　海洋環境創生」 (Marine Environment Creation Agency) for providing information and insight during our research for the utilization of algal mass. We would also like to thank TED speaker Dr. J. Trent from NASA for providing the inspiration and information for our research target.

References

[1] *Trent, Jonathan. "Jonathan Trent:Energy from Floating Algae Pods."* Jonathan Trent: Energy from Floating Algae Pods. TED, June 2012. Web. 01 Apr. 2014.

[2] *Dunbar, Brian. "NASA - OMEGA Project."* NASA. NASA, 11 Apr. 2012. Web. 02 Apr. 2014.

[3] *藻類に関する最新の動向・資料作成業務報告書 (Latest Trends & Data Business Report on the Utilization of Algae).* Tech. N.p.: 財団法人　九州環境管理協会 (Kyushu Environmental Management Association), 2011. Print.

[4] Figure 2. NASA OMEGA Project (2012) *OMEGA Photobioreactor Prototype* [Photograph; OMEGA Photobioreactor Prototype.] At: http://www.nasa.gov/centers/ames/research/OMEGA/news/imagegallery/index.html (Accessed on 04.05.14)

[5] "Clean Water Education Partnership." *Clean Water Education Partnership - Kids.* Clean Water Education Partnership, n.d. Web. 04 May 2014.

Do major construction projects endanger Stuttgart's mineral water resources?

Valerie Renger, Philipp Sauter

Dillmann-Gymnasium, Germany, *knaup@dillmann-gymnasium.de*

Abstract

Stuttgart has the second largest resources of mineral water Europe-wide. Due to the construction of a new underground main station in the centre of the city, the local mineral water resources are endangered. As a starting point our project will give general information about the mineral water resources and examine their quality with a chemical analysis to explain their importance for the region. Our focus is then going to move on to a critical review of the main station construction project (Stuttgart 21) and its consequences for the mineral water resources. By research into the data, expert input and modelling the situation, we want to evaluate the situation and ultimately find out if there is a modern and at the same time eco-friendly way to develop our city.

Keywords

Stuttgart 21, water quality, mineral water, ground water management

1 Introduction

Stuttgart, located in the south-west of Germany, is famous for its worldwide-known automobile industry, with companies like Porsche, Daimler or Bosch.

For many years the citizen of Stuttgart are polarised by the major construction project *Stuttgart 21,* a new, modern main station for the city to ensure efficient mobility for the future.

2 Status Quo

At the moment, Stuttgart has an overground terminus station. The new main station is planned as an underground through station. Unfortunately, this underground station coincides with the world's second largest resources of mineral water, located in the area where the new main station is built. More than twenty-two million litres of mineral water effervesce daily from nineteen springs all over the city.

Figure 1: Concept of Stuttgart 21

Furthermore, mineral water is used for three baths, creating a special experience, and can be drunk for free at public springs.

Figure 2: Public spring of mineral water

Since Stuttgart's citizen are proud of its resources as a cultural heritage, authorities allocated the spaces as an protected area.

To ensure the water's quality, the public institute *SES* regularly controls the amount of minerals and harmful substances [1]. Recent analyses showed high amounts of minerals like calcium, magnesium, sodium, potassium, chloride or sulfate. Figures show, that Stuttgart's mineral water has between ten and three hundred times more minerals than ordinary mineral water. In addition to that, not a single milligram of toxins or hydrocarbons could be detected [2].

Since the mineral water is distributed within the city, it has to be removed in order to build the underground station. Seventeen kilometres of blue pipes run through Stuttgart in order to remove ten billion litres of mineral water, which are lead into different underground areas and into the river Neckar [3].

3 Dangers & Solutions

Many citizens as well as some scientists fear the consequences of the current *ground water management*. As seen in other cities, e. g. Cologne, the reduction of ground water caused enormous impacts on nearby structure stability. In the worst case, according to some scientists, buildings could collapse or at least be damaged. Furthermore, the listed main station building in Stuttgart, which is part of the new remodel, already has cracks in its sandstone material, so the reduction of ground water could eventually lead to more damage as there already is [4].

Figure 3: Pipes for ground water management

Another point concerns Stuttgart's tourism. Because Stuttgart's mineral water resources are, as mentioned, unique in Central Europe, they attract tourists to visit the city. The springs are the heart of Stuttgart, on which its unique flora and fauna is based. Tourists come to Stuttgart in order to see the unique park *Schlossgarten*, which relies on the mineral water to survive [5].

In addition to that, the mineral water is essential for the citizens as well. They love their *Schlossgarten* as well as the three public baths, supplied with mineral water. Due to the high amount of minerals, no chlorine is necessary to ensure their water quality, leading to a unique experience especially for allergic citizens [6].

Nevertheless, the whole building process will be observed by the project management. Strict controls of the amount of pollution, potentially caused by the construction process, will be held to prevent the mineral water from pollution [7].

4 Conclusion

To conclude, a modernisation of Stuttgart's main station is important to ensure especially the city's economic status for the future. Notwithstanding, the environment, especially Stuttgart's unique mineral water resources, must not be harmed. Overall sustainable construction is necessary world-wide to preserve the world for future generations.

References

[1] http://www.stuttgart-stadtentwaesserung.de/fileadmin/user_upload/PDFs/SES_Zentrallabor_Flyer_kl.pdf
[2] http://www.stuttgart-stadtentwaesserung.de/fileadmin/user_upload/PDFs/Analysen/Jahresanalysen_2013.pdf
[3] http://www.stuttgart.de/grundwassermanagement
[4] http://www.geologie21.de/stuttgart-mineralwasser-mineralquellen/grundwassermanagement-stuttgart.html
[5] https://www.stuttgart.de/item/show/322481/1
[6] https://www.stuttgart.de/baeder/mineralbaeder
[7] http://www.bahnprojekt-stuttgart-lm.de/details/umwelt/wasser/grundwassermanagement/

Figures

[1] http://www.b4bschwaben.de/cms_media/module_img/218/109235_1_lightbox_Stuttgart_21_von_oben_Visualisierung_ingenhoven_architects_Quelle_das_neue_herz_europa s.jpg
[2] http://www.wuerttembergweb.de/fileadmin/processed/csm_Foto_1a_Berger_Sprudler_9a32c12bf0.jpg
[3] http://www.stuttgarter-zeitung.de/media.media.6f15a212-8ee2-4dd2-96cf-e76565c8d24f.normalized.jpeg

Disaster Prevention Wells

Riho Takeda, Mayo Sueta, Hiroshi Kawakatsu

Shibuya Senior High School, Japan, natsume@shibuya-shibuya-jh.ed.jp

Abstract

On the 11th of March 2011, Japan was hit by an undersea earthquake of magnitude 9.0 just off its coast, often referred to as the Great East Japan Earthquake. Many water facilities were damaged by the tsunami; at least 1.5 million households were reported to have been denied access to water supplies at the time. At such a time, wells became useful for providing drinking and domestic water for the citizens. It's resistance to earthquakes became an ideal water source during the emergency and the old method of getting water which has been fading out of the picture is now being looked at with a new light again. For example, in 3.11, 213 wells out of the 261 the Tohoku area that the organization was able to investigate did not receive any influence from the earthquake or the tsunami and was able to be used as usual and 34 of the remaining were fixed easily and were able to be used again. With the various technological advancements in the world, people often scoff at such things as wells but it has in fact proven itself to be a useful water supplier in emergencies.

Keywords

wells, natural disasters, water supply

Introduction

Imagine seeing only the stone foundations of your memory-filled family house that had once stood there among many others. Even after over three years after the massive earthquake hit the Tohoku area on March 11th 2011, many people still live in temporary housing and 267,000 citizens cannot go back to their hometowns due to the radiation from the Fukushima Nuclear Power Plant. On the other hand however, the situation in the Tohoku region has improved since the disaster thanks to both national and international help. Right after the earthquake and tsunami, the area had been destroyed completely; while aid was sent as soon as possible, the citizens of the Tohoku area suffered from harsh living conditions, including the lack of food and water. At that crucial time, wells played a large role in recovery efforts as it provided the citizens with drinkable and domestic water. Ever since its active part in supporting the citizens of Tohoku after the earthquake and tsunami, people have started to look at wells in a new light. Our research is centered on wells because although many people consider them to be an old and time-consuming way to get water, it has proved its strengths in emergencies and is too precious for its existence to dwindle away with time.

1 Content

1.1 The purpose of this investigation

The purpose of this investigation is to research on the use of wells as a source of water during emergency situations, especially during natural disasters. Through this investigation, we wish to learn about the damages and problems to waterworks caused by natural disasters by examining past data; the importance of water, including domestic water, during such situations; and how wells can supply safe water to people and support their lives.

1.2 Method of this investigation

We have researched and analyzed a vast number of information from various sources such as books and the internet. In detail, we have investigated into the use of wells in the past, such as the Great East Japan Earthquake of 2011 and the Great Hanshin Earthquake of 1995 summarized the reasons why wells are resistant to earthquakes and researched on the system of registered disaster prevention wells at present. Furthermore, we have actually visited disaster prevention wells in our region to experience what we have researched.

1.3 Results of the Investigation

From this investigation, we have learned that wells played an active role in supplying water to people after Great East Japan Earthquake and the Great Hanshin Earthquake because of its strong structure, and the demand for disaster prevention wells is increasing. Some administrations and organizations have taken in this demand and have constructed wells meant for disasters; some municipal authorities have adopted a system of registered disaster prevention wells. However, at the same time, we have also found that many challenges still remain in the wide diffusion of wells as a source of water during emergency situations.

2 Damage and Effects of Tohoku Earthquake and Tsunami

2.1 Overall Damage

At 14:46 of March 11th 2011, a massive undersea earthquake of magnitude 9.0, off the Pacific coast of the Tohoku region followed by an equally large-scale tsunami killed over 15,000 people and injured more than 6000 people together [1]. While around 800 people were either crushed or burned and died from internal injuries, the rest were swallowed by the waves of the tsunami and drowned [2]. Houses, cars, ships, schools, and most buildings were washed away and after the disaster, only the foundations were left to remind people of the buildings that once stood there. The earthquake and tsunami affected transportation as well; public transportation from other parts of Japan to the Tohoku region were temporarily unavailable. The incident of the Fukushima nuclear power station further worsened the situation. The tsunami destroyed the station; the radiation from the nuclear power station threatened the health of the citizens and they were soon evacuated out of their hometowns.

Of the remaining citizens who managed to stay alive, 1,331 died because of harsh living conditions such as lack of water, food and electricity [3].

Figure 1: The tsunami destroying the towns in the Tohoku region.

2.2 Damages to Waterworks

The total number of houses which suffered from the suspension of water was 2.57 million over 19 prefectures and 264 water supply businesses [4]. The causes for the suspension of water can be divided into two groups based on the prefectures that received the most damage (the suspension of water in over 300 thousand houses); in the Iwate, Miyagi and Fukushima prefectures, the main reason can be seen as damages caused directly by the earthquake and the tsunami, and in the Chiba and Ibaraki prefectures, the main cause was the liquefaction of the ground caused by the earthquake. Other causes were aftershocks after the main earthquake and power failure.

Approximately 2 million houses were restored two weeks after the earthquake, but damages endured for several more weeks due to aftershocks [5]. The restoration of all houses, excluding the 45 thousand houses in areas flooded by the tsunami, was completed 200 days later [6].

3 Preparation for the Next Natural Disaster

3.1 Possibility of Natural Disasters in the Future

There have been many predictions of more earthquakes to come since the March 11th earthquake in 2011. It is predicted that in the next three years, the possibility of an earthquake of over magnitude 8 hitting the Tokyo area was 99.9 percent [7]. The possibility of an earthquake hitting the Kanto region is higher than other regions as there are several plates under that area which overlap each other in a complicated way. There are 19 different types of earthquakes that could hit the Kanto region but experts have predicted that among them, an earthquake whose epicenter is directly below Tokyo is the one most likely to come in the near future [8]. The predicted number of casualties is 23,000 and it is said that around 600,000 houses will be destroyed [9]. Furthermore, the eruption of Mt. Fuji has been predicted based on the earthquake around the mountain area. Experts worry that the eruption of Mt. Fuji and a massive scale earthquake in the Kanto area will come around the same time, causing severe casualties.

3.2 The Necessity of Water during Emergency Situations

We all use around 3 liters per person per day to stay alive [10], and we can only live without water for around 3 days until we perish from thirst [11]. The average amount of domestic water that we use is 322 liters; 242 liters of the water is used at home at such places as the toilet (28 percent), bath (24 percent), cooking (23 percent) and laundry (17 percent) [12]. During the recovery efforts during the Tohoku region, most citizens were without water for both drinking and domestic needs. Firstly, lack of drinkable water can obviously lead to various problems of health. Furthermore, people are actually unaware of the fact that lack of domestic water can lead to health hazards and even death. According to the report on the Great Hanshin Earthquake disaster and the Niigata-Chuetsu Earthquake, the problem of lack of toilets led to deaths [13]. Because there were very few toilets in the devastated area, toilets were used until they overflowed. The possibility of death for the people who refrained from using toilets became higher than the people, who could use toilets in the same environment because people refrained from drinking water, leading to heart problems.

4 The Use of Wells as an Emergency Water Source

4.1 The Uses of Wells

Wells can be used as a method of securing water when water supplies are cut off due to natural disasters such as earthquakes or tsunamis. The water supplied by these wells can be used for several causes depending on its condition, such as drinking water, domestic water for washing, laundry, and the lavatory, and water to prevent secondary disasters such as fires. Although deep wells receive little influence from the surface, shallow wells are relatively easily affected; therefore, it is highly recommended that shallow wells be used for domestic purposes only, especially as the water can be contaminated by from surrounding drainpipes and bacteria such as colon bacilli after disasters. However, the installation of water purifying devices might improve this situation. Wells that are constructed specifically for providing water during emergency situations are called disaster prevention wells; these include wells that are managed by local governments or corporations, and also wells owned by individuals or private offices that are approved and registered by municipal authorities.

4.2 Great East Japan and Great Hanshin Earthquakes

The restoration of water supply facilities was greatly delayed after the Great Hanshin Earthquake in 1995. Two weeks after the disaster, over 490 thousand households were still suffering from the suspension of water, and some regions could not use tap water even two months after the disaster [14]. There are various reports of the use of wells during these times. According to these reports, well water was used for putting out fires by using buckets; food and brewery companies and ordinary households opened their wells to the public, where they were used for drinking and cookery; and water from both public and private wells were used widely for domestic purposes, such as for flushing the toilet and for laundry [15].

Wells also played a large role in the supplying of water recently, during the Great East Japan Earthquake in 2011. For example, in the Aoba ward of Sendai city, a private well supported the lives of the people of the local community while water was suspended for six days. The owner of the well and members of the community commented on the water from the well as the "water of life" and remarked that the well was useful for providing water to sustain their lives, such as for drinking, cookery, and baths [16]. Furthermore, following the active role of wells during the disaster, various facilities have adopted wells as an emergency source of water. One example is a disaster base hospital in Kochi prefecture, which needs 350L of water per day for the cleaning of surgical instruments and artificial dialysis, and has started on the construction of a well that can provide approximately 600L of water per day [17] Administrations have also taken notice of wells, and local self- governments are encouraging their people to register their wells as "Registered Disaster Prevention Wells" or are constructing public wells of their own.

5 The Structure of Wells

5.1 Categories of Wells

Wells can generally be distinguished into two types: the shallow well and deep well. The shallow well takes water from unconfined aquifers and is typically only around five to thirty meters deep, while the deep well takes water from confined aquifers and can be up to two hundred meters deep. Since the deep well takes water from underground water that flows under the first impermeable layer and receives little influence from the ground surface, its water temperature, quality, and quantity is said to be stable and safe. Therefore, water taken from deep wells has a tendency to be used for industrial purposes, as well as being used as a source of tap water. On the other hand, since the shallow well takes water from underground water that flows above the first impermeable layer, it is often influenced by the condition of the surface above, and can be contaminated relatively easily. Also, since shallow wells have a merit of being able to be constructed easily, it is common in ordinary households.

However, the quality of wells changes greatly according to the environment surrounding it.

5.2 Resistance to Earthquakes

After the Great East Japan Earthquake, the National Water Well Association of Japan carried out an investigation on the damage of deep wells caused by the disaster. Out of the 261 wells in the Tohoku area (Aomori, Iwate, Akita, Miyagi, Yamagata, and Fukushima prefectures) that the organization was able to investigate, it turned out that 213 (approximately 82% of the whole) did not receive any influence from the earthquake or the tsunami and was able to be used as usual; 34 wells (13%) did go through problems, but eventually recovered and became usable again; and 14 wells (approximately 5%) were no longer usable after the disaster [18]. Out of them, only 3 (about 1%) became broken because of the ground motion of the earthquake [19]. Also, experts note that ground liquefaction rarely causes any influence to wells, as the well's deep structure prevents any damage [20]. However, there have been reports of salt water brine damages to shallow wells by water supply businesses in 23 areas in 4 prefectures (Iwate, Miyagi, Ibaraki and Chiba) [21]. Therefore, it seems that although deep wells are highly resistant to natural disasters, shallow wells have a weakness toward tsunamis.

The materials used for the structure of deep wells explain its resistance. When copper pipe is used as casing pipe, it acts as steel pipe pile, which is resistant against vertical load. Since the condition of the structure would be almost the same as that of the base, the damages put on the wells can be limited [22].

5.3 Construction Fees and Term

The cost of constructing a well is never stable, as its price changes very easily. The construction fee of wells ranges from private wells at a reasonable price of two hundred thousand yen, which is equivalent to about two thousand US dollars, to public wells for industrial purposes, which costs approximately a million yen, which is about ten thousand US dollars. There is no guarantee that the cost will always be the same, as the cost is based on various factors such as its depth, the materials that are being used, and the hardness of the stratum. However, once the well is built, the well is almost free of charge due to the fact that it does not need to pump itself up to collect water. Although there is no clear price indicating how much it costs to fix a well, its price can be lowered by letting the professionals check on the well once in a while to prevent any damages.

The time needed to construct a well also depends on the factors listed above, but it is said to take approximately two weeks to dig up a hundred meters. At the same time, fixing a family well only takes about a few days.

6 Registered Disaster Prevention Wells

6.1 Description

Registration wells are wells owned by individuals or private offices that have been officially approved as disaster prevention wells by municipal authorities. Generally, local governments call for voluntary well owners who wish to register their wells, and examine them before recognizing them as official disaster prevention wells. By registering and spreading information about these wells, it becomes easier for members of the local community to secure water supply in times of need. The merits of having registration wells are that municipal authorities can secure a number of usable wells during emergencies with less cost compared to constructing new wells; wells can be expected to be a reliable, stable source that can compensate for the insufficiency of water after disasters; wells can provide domestic water and prevent the deterioration of public health; local wells can lessen the burden of having to carry large amounts of water when the nearest water supply base is distant; and since the system of registration wells relies entirely on the spirit of kindness and cooperation of the members of the community, a raise in the awareness of disaster prevention and regional cooperation can be expected.

6.2 Conditions for Registration

There are several conditions (differing by region) that have to be cleared for a well to be registered as disaster prevention well [23].

① That the well is located in the region controlled by the municipal authority

② That water from the well can be provided to others, free of charge, in the case of an emergency

③ That the well is in use at the present, and will continue to be used in the future

④ That the owner agrees to put up a visible sign displaying the existence of a registered disaster prevention well

⑤ That the owner agrees to make information about the well open to the public

⑥ That the well clears the water quality standards set by the municipal authority (Generally when the well is expected to become a source of drinking water)

Wells that are considered unsafe, such as wells that have been out of use for a long time or wells that are noticeably contaminated do not clear these conditions. Also, wells that are considered inconvenient for others to use, such as wells that are placed inside individual homes, also cannot be registered.

Figure 2: An example of a sign of registered disaster prevention well

6.3 Steps of Registration

The steps to register a well changes according to the region, but typically follows these steps [24].

① The owner of the well fills out a disaster prevention well registration sheet and hands it in or sends it to the disaster prevention section of the local municipal office. Registration sheets are usually available on official websites or at municipal offices. Some local governments accept applications by email or phone.

② The municipal authority examines the registration sheet and carries out water quality tests if necessary. If the well clears the conditions necessary for registration, the owner receives a document that officially acknowledges the well as a disaster prevention well.

③ The owner receives a sign to display the existence of a registration well so that members of the regional community are able to grasp the location of the well

The owner of the well can apply for the change or cancellation of registration through the same steps.

6.4 Warnings Considering the Usage of Registered Wells

① Registration wells can only be used in times of emergency, and are available only during emergency unless the owner states otherwise

② Users must remember that registration wells are supplied by voluntary spirit, and should follow the directions of the owner regarding the usage of the well

6.5 Present Situation of Registration Wells

We researched the present situation of registered disaster prevention wells in the 23 wards of Tokyo.

Table 1: Present Situation of Registered Wells

Name of Ward	Number of Wells	Remarks
Chiyoda	22[25]	
Shinjuku	179[26]	Shallow wells
Taito	9/33/33[27]	Deep, public/ deep, private/ shallow
Toshima	442[28]	Mainly for preventing secondary disasters
Kita	17[29]	Deep
Arakawa	39/4[30]	Public/ Private
Nerima	2/ 500[31]	Deep/Miniature Type
Koto		Not suitable[32]
Bunkyo, Koto, Meguro, Oota, Setagaya, Nakano, Suginami, Adachi, Katsushika	Number Unknown	
Chuo, Minato, Sumida, Shinagawa, Shibuya, Itabashi, Edogawa	No Information Available	

Number Unknown: Registration of wells is available, but the direct number of wells was not available

7 Further Improvements to the Use of Disaster Prevention Wells

7.1 Registration Wells

7.1.1 Spreading Information About Registered Wells

Out of the 23 wards of Tokyo that we researched, 7 wards had no information about the registration of natural disaster prevention wells on their official websites. Furthermore, 8 had information about the registration of wells but did not have detailed information about the number or the location of the wells. It is necessary to spread information about natural disaster prevention wells to actually put them in use, as the purpose of registration wells is to provide a stable source of water to members of the local community during emergencies. Information about registration wells should actively be given to the public through media such as official websites, neighbourhood associations, public magazines, posters, and disaster prevention maps. Cooperating with regional disaster prevention organizations to spread information and to enforce emergency drills might also be effective. This may not only make the system of disaster registration wells more effective, but would also encourage individuals to register their wells or to restore old wells.

7.1.2 Clarifying Conditions of Registered Wells

The conditions and rules about registration wells are unclear in some areas; there is a need to arrange and clarify them. One example is the valid registration period. Some municipal authorities have set a valid registration period of 1~5 years, or have enforced water quality tests once every 1~2 years to check the well's condition [33]. However, some have not set a registration period and there is no method to check the well's recent condition. This might result in cases where the owner of wells do not submit documents concerning changes concerning the well or the discontinuance of its use, leading to the decline of the reliability of information given to the public. Therefore, there is a need to set a valid time period or regular tests and checks to keep the system effective.

Another example is the problem of the lack of assistance from municipal authorities. Although some local governments provide assistance payment for the construction or repairing of wells, or provide water quality tests every few years, many leave expenses and tests to the owner's responsibility. By providing some form of assistance, local governments can lessen the burden of owners and can also encourage people to repair old wells or construct new ones.

7.2 The Construction of Public Wells

The system of registered disaster prevention wells relies entirely on the voluntary spirit of the members of the local property; therefore, there might be a lack of usable wells in certain areas. Accordingly, municipal authorities may need to consider a project of building disaster prevention wells

themselves in main evacuation shelters and public facilities. One example of a municipal authority which has already started this project is the Nerima Prefecture in Tokyo, which has established deep wells around 100m deep in two areas, shallow wells around 25m deep in several public schools, and miniature shallow wells around 9m deep in more than 500 areas in the region [34]. The deep wells are normally used for drinking water, but have emergency power generators, taps and hoses attached to it to use during times of disaster. The shallow wells located in public schools can be used for domestic water, and has an electrically- powered pump attached. The miniature wells are hand- powered, and can be used to prevent secondary disasters and for domestic water during emergency situations.

By carrying out programs such as this, municipal authorities can ensure that everyone in the local community has a disaster prevention well within close proximity to secure a supply of water during emergencies.

7.3 Other Merits of Disaster Prevention Wells

There are several uses for disaster prevention wells other than the providing of water during emergency situations. Wells should be used regularly, as the daily use of wells keeps the quantity and quality of the water stable and safe.

Firstly, water from wells in individual wells can be used in daily life as domestic water; for example, laundry, gardening, and washing the car. By using well water, people can cut down on water bills, as drawing water from the well is generally free after its main construction. It is also possible to use well water for drinking through regular water quality tests and the attachment of water purifying devices.

Secondly, disaster prevention wells can be used for education through constructing them in public facilities such as schools and parks. An example of this is the disaster prevention school well education project in Kyoto. 46 schools in Kyoto prefecture have participated in this program and have constructed hand- pumped wells within the school grounds [35]. The water from these wells is used on a daily basis for watering plants and the school playground. Furthermore, the wells serve as teaching material for science classes and environmental education on water resources. It can also serve as a chance for young people to experience people's lives before the development of recent water technology.

Thirdly, public wells can serve as a place for communication between members of the regional community. During the Edo period, the maintenance of sewerage systems progressed, and shared wells were established in many regions. Since there were no tap water back then, people used water from these wells for their daily lives. It was not unusual for members of the local community to chat with each other while waiting in line to draw water; "Idobatakaigi", meaning "idle gossip" is a widely used expression that came from this scene. Kokubunji City has brought this scene back to the present day by holding "Idobatakaigi" at disaster prevention wells for approximately an hour every month [36]. At these meetings, the maintenance of hand pumps and simple water quality tests are held, and people can learn not only about wells and underground water, but can also share information about disaster prevention and local community news with neighbours while drawing water from the well.

8   Conclusion

In conclusion, despite the concerns people may have with wells, it is definitely true that wells are capable of playing an important role in helping people have access to water in case of emergencies. People may have seen the wells as old-fashioned and time consuming prior to the Great East Japan Earthquake. However, after the incident, many people, including the citizens of the areas affected by the earthquake and tsunami, have begun to realize that wells have played a large role in securing water. Moreover, the resistance of wells in earthquakes is very helpful when it comes to any type of natural disaster. We can benefit from this disaster by making the Great East Japan Earthquake a chance to learn that old technologies such as wells can benefit and support our lives in emergency.

Acknowledgements

We would like to thank the teachers who supported us during our research, including Mr. Nancoo for his assistance in choosing the members of this project. We would not be here without his help. Most importantly, however, we would like to express our deep gratitude towards Ms. Natsume who repeatedly sent us emails concerning the Water is Life project, took care of the paper work. Thank you for supporting our work and research.

References

[1] "Damage Caused by Japan Earthquake, Tsunami, Nuclear Crisis 2011." Liberal Sprinkles. March 16th 2011. March 17th 2014 < http://liberalsprinkles.blogspot.jp/2011/03/damage-caused-by-japan-earthquake.html>

[2,3]Abubakar, Nasiri. "One year after the Great East Japan Earthquake, Tsunami." Sunday Trust. 15th April 2012. March 17th 2014 <http://www.sundaytrust.com.ng/index.php/feature/2996-one-year-after-the-great-east-japan-earthquake-tsunami>

[4, 21]"A Summary of the Damages Caused to Waterworks Facilities by the Great East Japan Earthquake." Ministry of Health, Labour and Welfare. 29 March 2014. < http://www.mhlw.go.jp/stf/shingi/2r9852000002qek5-att/2r9852000002qep4.pdf>

[5]"Suspension of Water: Still Over 40 thousand Households not Restored." Tokyo Shimbun. 11 December 2012. 30 March 2014. < http://www.tokyo-np.co.jp/article/feature/tohokujisin/nattokuqa/list/CK20121211102000145.html>

[6]Obara, Takuro. "An Analysis of the Damages to Waterwork Facilities by the Great East Japan Earthquake." Association of Water and Sewage Works Consultants Japan. 30 March 2014.< http://www.suikon.or.jp/report/pdf/27/h24_13-18.pdf>

[7-9] "Prediction of the casualties in future Earthquakes in the Tokyo area." Asahi Newspaper Digital. March 28th 2014 < http://www.asahi.com/special/syutochoka/ >

[10, 12] "Substitution for Water Pipes." Ikeura Bousai net. March 21st 2014 < http://www.ai21.net/bousai/sonae/infra/>

[11] "How Long Can A Person Live Without Water." Livescience. November 30th 2012. March 22nd 2014 < http://www.livescience.com/32320-how-long-can-a-person-survive-without-water.html >

[13] "Mission- Chain of Greencity." Kakogawa Greencity Official Website. 29th March 2014. <http://www.japanriver.or.jp/taisyo/oubo_jyusyou/jyusyou_katudou/no9/no9_pdf/kakogawa.pdf>

[14] "Wells Noticed After Disaster." JCA- NET. 30 March 2014. < http://www.jca.apc.org/water-w/kobeidokyokun.html>

[15]"The Use of Underground Water During Earthquake Disasters." Ministry of Health, Labour and Welfare. 30 March 2014. < http://www.mhlw.go.jp/topics/bukyoku/kenkou/suido/topics/suijunkan/dl/051114-2d4b.pdf>

[16,17,20]"Wells Noticed after Disaster." NHK Online. 4 October 2013. 30 March 2014. < http://www9.nhk.or.jp/nw9/marugoto/2013/10/1004.html>

[18,19] "A Report of Damages to Wells by the Great East Japan Earthquake." National Water Well Association of Japan. July 2012. 30 March 2014. <http://www.sakusei.or.jp/ido_report.pdf>

"For People who Wish to Construct Disaster Prevention Wells." Subsurface Investigation Office. 30 March 2014. < http://www.jiban-chosa.co.jp/topics1-4.html>

[23,24,33]"Guideline to the Introduction of the Registered Disaster Prevention Well System." Shiga Prefecture Official Website. January 2013. 29th March 2014. < http://www.pref.shiga.lg.jp/e/seikatsu/suidou/files/guidelineh2501.pdf>

[25] "Other Disaster Prevention Policies of Chiyoda Ward." Chiyoda City Official Website. 20 February 2014. 30 March 2014. http://www.city.chiyoda.lg.jp/koho/kurashi/bosai/sonota/shinsai.html

[26] "Shinjuku Ward Disaster Prevention Program." Shinjuku City Official Website. 2008. 30 March 2014. http://www.city.shinjuku.lg.jp/content/000038320.pdf

[27] "The Voices of the Community." Taito City Official Website. October 2012. 30 March 2014. http://www.city.taito.lg.jp/kouho/kouchou/teian/voice_category/category/2012_10/2012_10_06.html

[28] "The Securing of Fire Fighting Water." Toshima city Official Website. 10 July 2008. 30 March 2014. https://www.city.toshima.lg.jp/kusei/kusei/bousai_machizukuri/chiikibousai/000672.html

[29] "About Disaster Prevention Facilities (Deep Wells)." City of Kita Official Website. 1 April 2013. 30 March 2014. http://www.city.kita.tokyo.jp/docs/service/167/016789.htm

[30] "Disaster Prevention Wells." Arakawa City Official Website. 21 June 2012. 30 March 2014. http://www.city.arakawa.tokyo.jp/kurashi/bosaibohan/bichiku/ido.html

[31,34]"Disaster Prevention Wells, School Disaster Prevention Wells, and Miniature Disaster Prevention Wells." Nerima City Office Official Website. 1st February 2013. 29th March 2014. < http://www.pref.shiga.lg.jp/e/seikatsu/suidou/files/guideline h2501.pdf>

[32] "Opinions of the Community." Koto City Official Website. 19 January 2012. 30 March 2014. <http://www.city.koto.lg.jp/pub/faq/faq_detail.php?fid=737 6>

[35]"Disaster Prevention School Well Project." Kyoto City Official Website. 4th April 2011. 29th March 2014. <http://www.city.kyoto.lg.jp/kyoiku/page/0000096798.html >

[36]"Mukashi no Ido." Kokubunji City Official Website. 25th July 2012. 29th March 2014. <https://www.city.kokubunji.tokyo.jp/anzen/5967/006496.h tml>

[37] "The Construction of Disaster Prevention Wells." Kakogawa Greencity Official Website. 29th March 2014. < http://www.greencity.sakura.ne.jp/greencity_bousaikai/idoh ori_daisakusen/bousai_ido_1.htm>

[38] Sekine, Ryohei. "A Report of Damages of the 3.11 Great Earthquake and Tsunami with the Photograph and the Movie in Fukushima and Miyagi Prefecture." The 2011 East Japan Earthquake Bulletin of the Tohoku Geographical Association. 1 May 2011. 29 March 2014. < http://tohokugeo.jp/disaster/articles/e-contents15.html>

[39] "The Estimation of Damage by Projected Tokyo Earthquake." The Asahi Shimbun. 29 March 2014. < http://www.asahi.com/special/syutochoka/>Mayo Clinic Staff.

[40] "Water: How Much Should You Drink Every Day?" Mayo Clinic. 12 October 2011. 29 March 2014. < http://www.mayoclinic.org/healthy-living/nutrition-and-healthy-eating/in-depth/water/art-20044256>

[41] Binns, Corey. "How Long Can a Person Survive Without Water?". Live Science. 30 November 2012. 29 March 2014. < http://www.livescience.com/32320-how-long-can-a-person-survive-without-water.html>

[42] Perlman, Howard. "Domestic Water Use." U.S. Geological Survey. 17 March 2014. 29 March 2014. < http://water.usgs.gov/edu/wudo.html>

[43] Oskin, Becky. "Japan Earthquake & Tsunami of 2011: Facts and Information." Live Science. 22 August 2013. 30 March 2014. < http://www.livescience.com/39110-japan-2011-earthquake-tsunami-facts.html>

Solutions to floods in Copenhagen which improve the appeal and image of the city

Gustav Høeg Andersen, Amalie Lohse Røntorp, Marcus Alexander Skytt

Vordingborg Gymnasium & HF, Denmark, lt@vordingborg-gym.dk

This study focuses on solutions to floods in Copenhagen, which improves the appeal and image of the city. Based on collection of information about possible solutions to floods in the city, experiences from other cities, and socioeconomic conditions in different neighbourhoods, we suggest a solution which secures a great area of the city, which is highly vulnerable to floods. The solution include reopening of the Ladegårds Stream and construction of a SMART tunnel below, lowering of the water level in the lake where the Ladegårds Stream ends, and construction of an open canal from the lake to the harbour. This solution will handle sufficient amounts of water during floods, as well as providing the city with a highly visible project, which also contribute to less congestion, a more integrated system of green areas, and increased integration of different areas.

Keywords

Copenhagen, floods, image, stream- tunnel-lake-canal-solution

Introduction

The heavy rains which Copenhagen has experienced in later years were not caused by the polar front where dynamic low pressures are formed and hot air rises above colder air, thus forming precipitation. Instead they were caused by further local warming of already hot and humid air, resulting in the heated air to rise, causing the water vapour to condense and precipitate. These events happen during summer and causes heavy rain because of the great amount of water vapour which the hot air contains.

Since the beginning of the 1990s, Copenhagen has been experiencing changes due to globalization and transition from industrial to knowledge society. The national and city authorities have responded to the new challenges and opportunities by investing in infrastructure, architecture, culture, the natural environment and an image, which people can relate to the city. In Copenhagen there has been great focus on integrating the natural environment (both green and blue areas) in the development of the city in order to make the city more attractive, liveable and sustainable, thus contributing to the image of the city as a green, sustainable and liveable city. Another aspect of a sustainable and liveable city is a high degree of equality in living standards between different neighbourhoods of the city.

Since floods have become more frequent in Copenhagen, a lot of initiatives have been taken to diminish future damage costs by floods. A lot of different solutions have been proposed, but one is not necessarily better than the other.

Adjusting Copenhagen to incidents of heavy rain is still a new process. However, general principles for solutions have been developed, but knowledge about specific solutions and their effect is still confined. Nevertheless, various solutions have already been suggested for different areas of the city.

Purpose

The purpose of this study is to contribute to the discussion of the most suitable solutions to higher frequencies of floods in Copenhagen with regard to both number and magnitude of floods, and how these solutions can be combined with the appeal and image of the city.

Method

The research on solutions to floods caused by heavy rain in Copenhagen is still new. However, a lot of research has already been done.

The research includes predictions of future climate and its potential effects regarding floods in different areas of the city [1, 2, 3], general assessment of suitable solutions in different types of areas [1, 2, 3], and specific solutions in different areas [4]. The research has been based on predictions of consequences based on IPCC's scenarios for future climate, the URBAN MIKE method to predict surface water flows [1, 2], and different tests of various solutions.

During our research we have found several studies already conducted. However, these studies have mostly focused on economically responsible solutions to reducing damages of floods. More indirect effects of the solutions, such as appeal of the city and contribution to the image, are also mentioned. Nevertheless, these effects are more difficult to assess.

Thus, our method has been to collect information regarding solutions which minimize floods caused by heavy rain in Copenhagen as well as information about Copenhagen's image and strategy regarding the appeal of the city. Furthermore, we have collected information concerning the effects of implemented solutions elsewhere. Collection of information about socioeconomic conditions in different neighbourhoods has also been conducted. Using all this information, we come up with a suggestion for the best solution. We do not suggest solutions for every street in Copenhagen. Instead we focus on the city as a whole, as well as specific areas of the city.

Results

The topography of Copenhagen shows a general inclination towards the east. This causes surface water to run in an eastward direction, except for the many instances where local depressions and the built environment have changed the flow direction of the surface water.

The reports which have been conducted suggest some general principles in order to minimize or avoid floods [1, 2, 3, 4]. These include retaining and/or slowing down the surface water in higher altitude areas, and transportation of surface water through big corridors in inclined and lower altitude areas. Areas adjacent to these water transporting corridors should lead water towards the corridors as well as absorbing it and slowing it down.

A number of streets have been pointed out as water corridors which should transport the surface water towards the harbour (in some cases via the inner lakes of the city).

Regarding the image of Copenhagen and its appeal to tourists, companies, and potential newcomers, we do support an emphasis on surface solutions which add blue and green areas to the city. Especially, solutions which contribute to connecting the various green/blue areas of the city, thus contributing to an integrated system of green and blue areas, should be prioritized. This could improve conditions for bicycles and pedestrians, thus promoting these means of transportation. Together, the mere existence of an interconnected system of green/blue areas, and the (expected) resulting increase in transportation by foot and bicycle will contribute to the image of the city.

We also emphasize the importance of huge visible projects. An initiative that, if realized, would be a major project is Åboulevarden and reopening of the Ladegårds Stream which run in a pipe under Åboulevarden. Åboulevarden which is a major street running through Copenhagen has been pointed out as one of the water corridors. Attention has therefore been directed towards this major street in order to prevent floods. However, various solutions have been suggested regarding Åboulevarden and its surrounding areas [3, 4, 5, 6].

Earlier, the Ladegårds Stream used to flow through the streets of Copenhagen, but due to a more congested traffic the Ladegårds Stream was brought under the earth and now flows through a pipe. One idea is to reopen the stream and let it flow through Copenhagen again. Two tunnels for cars should then be built under the river. In case of a major flood, one of the two pipes in the tunnel below would be able to close for cars and receive excessive rainwater. This would ease the pressure on the Ladegårds Stream. The other pipe would then temporary be used for traffic in both directions. This kind of tunnel system is called Smart (Stormwater Management and Road Tunnel) and has also been used and proved viable in Kuala Lumpur [6].

Similar projects in other cities have shown that there is a lot of potential with a stream like the Ladegårds Stream.

We know from Århus, the second largest city of Denmark (approx. 250.000 inhabitants) that the reopening of the urban stream created a new urban space full of life [6].

In Malmø, Sweden, the socially deprived neighbourhood of Augustenborg has experienced urban renewal based on a series of surface solutions to floods, including green roofs, canals, infiltration, and retention. This has resulted in more happy and proud residents, less crime, and greater turnout among the citizens. Additionally, as a part of the project, a Scandinavian Green Roof Institute was established [7].

The reopening of the Cheonggyecheon Stream in Seoul has proven to be a great solution. Benefits for the city include protection against heavy rain statistically only occurring once every 200 years, reduced small particle air pollution, increasing use of public transport, relative increase in property prices compared to the rest of the city, increased number of workers, businesses, visitors and tourists. 75% of the demolished concrete that used to be a highway on top of the stream was reused. In addition, the temperature reduced as much as 3.6°C in the areas around the stream compared to other areas in the city of Seoul [8].

Thus, experiences from Kuala Lumpur, Århus, Malmø, and Seoul suggest that a solution combining reopening of the stream and construction of a SMART tunnel beneath could combine various aims. However, we suggest that some conditions specific to the area around Åboulevarden further support the solution mentioned above. Åboulevarden runs on the border between Københavns Kommune and Frederiksberg Kommune, two municipalities in Copenhagen. In general the areas north of Åboulevarden are doing significantly worse than the areas located on the southern side, regarding unemployment and public benefits (see figures 1 and 2 below).

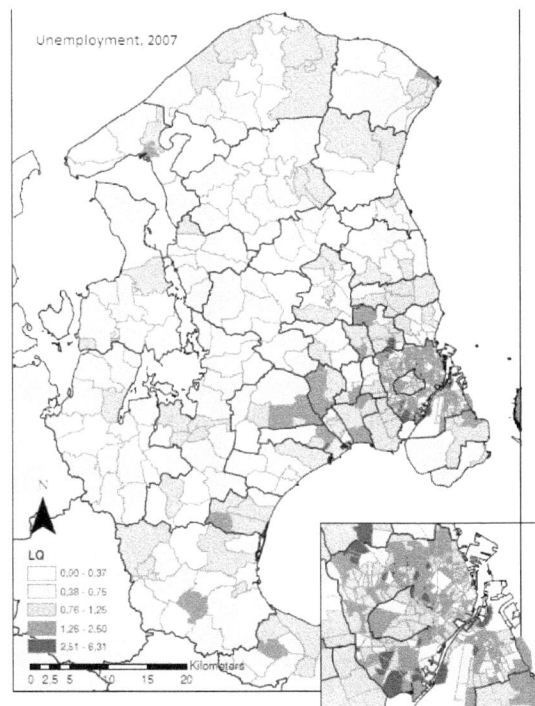

Figure 1: Unemployment in Copenhagen, 2007. LQ=1 indicates that an area has the same percentage of unemployed people as the whole area covered by the figure. LQ > 1 indicates that the percentage of unemployed people in an area is higher than the percentage of the whole area covered by the figure. We have added a green and a purple line to the small square in the figure. The green line represents the location of the Ladegårds Stream and the purple line represents the location of the suggested stream running from Skt. Jørgens Sø to the harbour.

Source: [9]

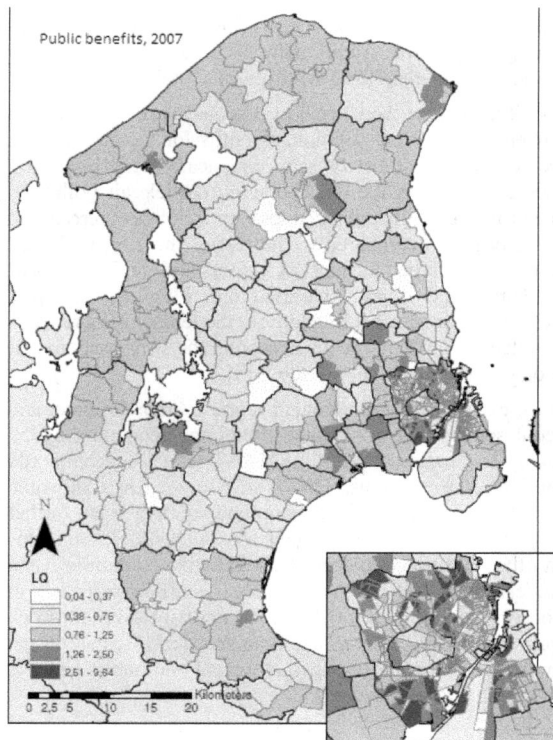

Figure 2: Public benefits in Copenhagen, 2007. LQ=1 indicates that an area has the same percentage of people receiving public benefits as the whole area covered by the figure. LQ > 1 indicates that the percentage of people receiving public benefits in an area is higher than the percentage of the whole area covered by the figure. We have added a green and a purple line to the small square in the figure. The green line represents the location of the Ladegårds Stream and the purple line represents the location of the suggested stream running from Skt. Jørgens Sø to the harbor.

Source: [9]

We suggest that substituting the big road with a green corridor and a stream will remove the physical barrier between the different areas and create a nice urban space where people from different areas can meet. However, it is also likely that the project will cause increased property prices, which was seen in Seoul. This can cause gentrification, thus changing the character of some of the areas mainly located to the north. This can somehow make the city more appealing, but also have an excluding effect and make the city more uniform.

Regarding traffic, a research project conducted by Rambøll [5] suggests that the tunnel will increase the traffic in Åboulevarden, but also decrease the traffic in adjacent streets, some of which are now often congested. Because of the great amount of pollution caused by congested traffic, we suggest that the tunnel solution is a good alternative even though the traffic on Åboulevarden will increase since it will cause less total congestion. Furthermore the tunnel will be able to prevent some of the pollution (not CO_2) entering the atmosphere and the ground.

The solution also helps connecting green areas of the city, thus improving conditions for greater biodiversity and more transportation by foot and bicycle.

However, a major problem with the solution is the building period which will cause huge traffic problems due to the fact that Åboulevarden is a road with great traffic. The project would definitely impede traffic for a long period.

Thus, we support the already suggested solution implying reopening of the Ladegårds Stream and placing the existing road under the reopened stream. This solution will transport surface water towards the lakes in the city center (and further on to the harbour), both during normal and intense rainfall (see figure 3 below).

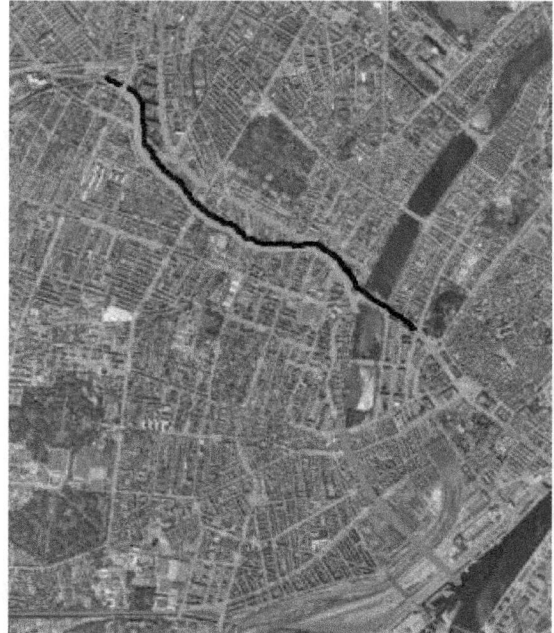

Figure 3: Blue arrows show direction of surface water. The green line represents the location of the reopened Ladegårds Stream. The black line represents the location of the suggested tunnel. The purple line represents the location of the suggested stream running from Skt. Jørgens Sø to the harbour.

We have added the green, black, and purple lines to a map, which have been constructed for another similar solution. However, the location and direction of the blue arrows are the same for the two solutions.

Source: [3]

If the project is to be realized, we should get inspiration from the Cheonggyecheon Stream in Seoul. Fountains, waterfalls and steppingstones surrounded by vegetation all contribute to the idyllic picture and create great areas for picnics, walks, tourism etc. However, we also know from Augustenborg in Malmø that inclusion of the residents in the design of the area will increase their engagement and sense of belonging to the place. The solutions should therefore be adjusted to the character of the different areas and the desires of the residents. Perhaps a playground including a skating area which is able to retain water could be an idea.

Furthermore, we suggest that the tunnel should continue under the inner lakes while the stream should enter Skt. Jørgens Sø, which is one of the inner lakes. Allowing the tunnel to continue under the lakes will make the lake-area more appealing. This appeal could be further improved by lowering the water level in Skt. Jørgens Sø, which is suggested for another solution, which does not include a tunnel and a stream in Åboulevarden [4]. Lowering the water level in Skt. Jørgens Sø could change the appearance of the lake considerably, including the establishment of a beach (see figure 4 below). This would make the area even more attractive than it is today.

Figure 4: Skt. Jørgens Sø as it could appear after lowering the water level.

Source: [4]

The research which have been conducted also suggests that water from the lakes and the areas facing the western sides of the southern lakes should be transported in a corridor towards the harbour in the south, which will also receive water from parts of Vesterbro, which are also highly vulnerable to heavy rain (see figure 3 and 5). We suggest that this corridor should be constructed as a surface solution wherever it is possible. The more narrow streets which this corridor will pass will not allow the same kind of solutions to be applied as in Åboulevarden. However, a solution could be to remove motorized traffic from the streets where the stream is running.

Figure 5 shows the vulnerability to floods of different parts of the city, and the solution suggested in this report.

Conclusion

Relying on experiences from other places and a number of analysis already conducted, we suggest that the Ladegårds Stream should be reopened and a SMART tunnel should be constructed below the stream. A SMART tunnel has already proved viable in Kuala Lumpur, and the stream above which will also be able to transport great amount of water during heavy rain will make the solution even more robust. This robustness is further enhanced by the absorbing capacity created by the lowering of the water table of the lake into which the Ladegårds Stream is running and the draining of the lake through a water corridor towards the harbour. Furthermore, the solution will contribute to less congestion, a more integrated system of green areas, increased integration of different areas as well as providing the city with a big visible project which will attract attention from tourists, potential newcomers, and people dealing with solutions to preventing floods caused by heavy rain.

A number of aspects of the suggested solution need further investigation:

Will the quality of the surface water running to the lakes and the harbour be acceptable; what effect will the future metro ring line have on transport patterns in the city; should the tunnel be extended under the harbour to Amager, thus freeing the whole city center from the big traffic corridor.

Further investigations and experiences in the future will improve our knowledge about solutions to heavy rain. However, we believe that the suggested solution is capable of securing a big area of the city, which is highly vulnerable to floods as well as improving the image and appeal of the city.

Figure 5: The colours indicate calculated floods (in meters) during heavy rain which statistically only occurs every 100 years, if no solutions are applied in order to prevent floods.

Source: [4]

Acknowledgements:

We would like to thank Martin Drews, Per Skougaard Kaspersen and Simon Bolwig from the Technical University of Denmark for discussing the project with us. Furthermore, we would like to thank our teachers Anne-Mette Christiansen and Lars Rostgaard Toft for their guidance throughout the project.

References

[1] Københavns Kommune (2011).
Københavns Klimatilpasningsplan. Formula.
pp. 1-99.
http://www.kk.dk/da/om-kommunen/indsatsomraader-og-politikker/natur-miljoe-og-affald/klima/klimatilpasning

[2] Frederiksberg Kommune (2012).
Klimatilpasningsplan 2012. pp. 1-66.
http://www.frederiksberg.dk/Politik-og-demokrati/Politikker-og-
strategier/Miljoe-klima-og-
affald.aspx#B076A2A0F511441EB347FC42CDF2AAF2

[3] Københavns Kommune (2012). *Skybrudsplan 2012.* pp. 1-32.
http://www.kk.dk/da/Om-kommunen/Indsatsomraader-og-
politikker/Publikationer.aspx?mode=detalje&id=1018

[4] Københavns og Frederiksberg Kommuner (2013).
Konkretisering af Skybrudsplan – Ladegårdså, Frederiksberg Øst og Vesterbro. pp. 1-14.
https://subsite.kk.dk/~/media/0ABB6FADE4124980B240746BF2D5F5
51.ashx

[5] Jensen, A.J. (2014). *Åbn Åen*
http://www.ladegaardsaaen.dk/

[6] Madsen, H. (2012). *Kampen om Åen*
http://www.magasinetkbh.dk/indhold/ladeg%C3%A5rds%C3%A5en

[7] Miljøministeriet / Naturstyrelsen (2014).
Augustenborg: Et ta' selv bord af tilpasnings-løsninger
http://www.klimatilpasning.dk/aktuelt/cases/items/augustenborg-et-ta-
selv-bord-af-tilpasningsloesninger.aspx

[8] Landscape Architecture Foundation (2014).
Cheonggyecheon Stream Restoration Project
http://www.lafoundation.org/research/landscape-performance-
series/case-studies/case-study/382/

[9] Samson, J. (2009). *Social segregation i danske byer.*
Institut for Geografi og Geologi, Københavns Universitet. pp. 4-39.

Examining the Effectiveness of RI Water Week as an Advocacy Event

Josiah Kek

Raffles Institution, Raffles Institution Lane, Singapore 575954

Abstract

RI Water Week was a week-long campaign in RI(JC) to raise awareness of water issues among RI students. These issues consisted of the global water crisis, solutions to water scarcity and the importance of saving water. Various methods of outreach were employed, including exhibitions and a symbolic walk to fetch water.

This research studies the effectiveness of RI Water Week as an advocacy event. Quantitative measures and qualitative responses from surveys and interviews will be examined. These datum will be used to determine whether RI Water Week was effective in meeting short-term and long-term outcomes.

Keywords

Advocacy, water, students

1. Background

1.1 RI Water Week

RI Water Week was an week-long campaign in RI (JC) to raise awareness of water issues among RI students. It took place from 21st April to 26th April 2014 and was organised by a team of 5 students.

The campaign had three aims:

- To make Rafflesians aware of water shortage and its importance
- To raise funds for water security projects in Ethiopia.
- To teach water conservation habits to Rafflesians

A wide range of methods was used to achieve these general aims. This included information booths, electronic posters and the inaugural Walk For Water. Each method had its own mini-aim(s).

1.2 Thirst in Ethiopia Booth

Mini-Aim(s): (i) To raise awareness of water scarcity in Ethiopia

Details: A booth was set up with information on water scarcity in Ethiopia. This included facts on Ethiopia's dry climate and its seasonal droughts. The 2 key problems facing Ethiopians (dirty water and water shortage) were also described with the use of statistics.

Photos of Ethiopia during drought were also on display as shown in Figure 1.

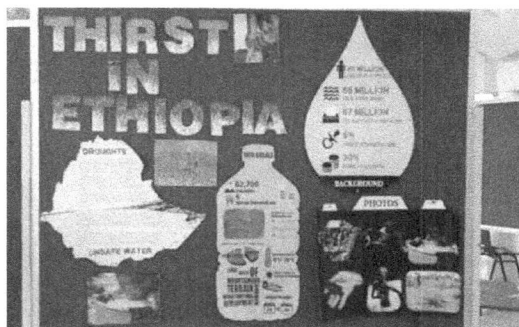

Figure 1: *Thirst in Ethiopia* booth highlighting the issue of water scarcity in Ethiopia

1.3 H^{2}OPE Donation Drive

Mini-Aim(s): To raise funds for water projects in Ethiopia

To increase awareness of measures to reduce water scarcity.

Details: A donation booth was set up in the Canteen throughout the week. Information was provided on three water projects in Ethiopia, namely spring construction, reservoir-digging and hygiene classes. These projects were labelled "1", "2" and "3", and information on each of them was featured. After reading the information, students could choose one project and donate to the glass jar labelled with its number.

Students who donated were given a special ribbon in appreciation (Figure 2).

Figure 2: H^{2}OPE Donation Drive set up to inform students and to provide an avenue for contribution for the water projects in Ethiopia.

1.4 Save That Drop Movement

Mini-Aim(s):

   (i)   To raise awareness of concrete ways to save water

   (ii)   To make students commit to water-saving measures

Details: A water-saving pledge movement was organised in the Canteen. Students were asked to make their pledges (Figure 3) on small whiteboards. They rotated the colourful wheels on this whiteboard to decide on whether they wish to save water in school or at home, and on which measure they are taking.

Figure 3: Pledges were made by students to save water

1.5 Water Issues Talk

Mini-Aim (s):

(i) To raise awareness of threats to Singapore's water security.

(ii) To make students understand the long-term impacts of water conservation.

Details: A talk by Mr Chew Men Leong, CEO of the Public Utilities Board, was held for all Year 5 students (Figure 4). His topic was "The Importance of Saving Water in the 21st Century". Mr Chew to spoke about the recent dry spell in Singapore and the need for all Singaporeans to be conscious of their water usage.

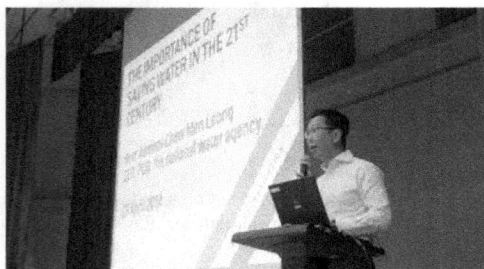

Figure 4: Mr Chew Men Leong, CEO of the Public Utilities Board giving a talk on The Importance of Saving Water in the 21st Century

1.6 Thirsty Tilaya App

Mini-Aim (s): (i) To allow students to gain knowledge about water scarcity in a fun way.

Details: *Thirsty Tilaya* was a game app (Figure 5) that was launched to the students. In the game, students would play the role of an Ethiopian child in search of water. Jumping from droplet to droplet, it was just a matter of time before they ran out of water. After the end of each round, a fact about water scarcity appears on the screen.

The game was designed by one of our team members for iOS/Android use.

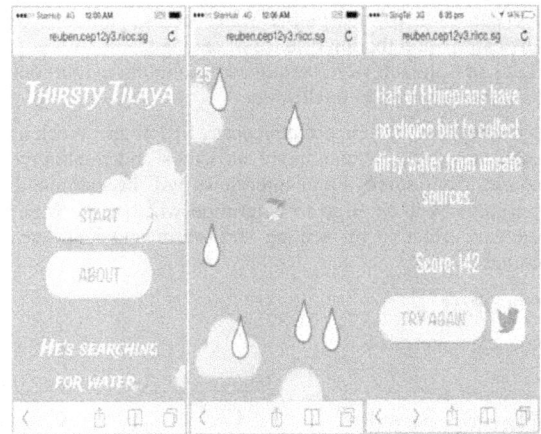

Figure 5: Thirsty Tilaya App provides an interactive interface for students to appreciate the problems faced by individuals in water scarced countries.

1.7 Walk For Water

Mini-Aim(s):

(i) To increase empathy towards the struggles of people who walk to fetch water.

(ii) To remind students not to take clean water for granted.

Details: Walk For Water was a 5km walk that simulated the long walk to fetch water taken by millions of people daily. Every two Walkers were given a 10L bottle at the start of the walk. They would fill up the bottle with water at MacRitchie Reservoir and embark on a 5km walk with the heavy load. Figure 6 showcases the participants who took part in this activity.

Figure 6: Group picture of the students who volunteered to take part in the Walk For Water campaign.

2. Introduction

RI Water Week was a student-led campaign on water issues that took place in RI (JC). It experimented with new advocacy methods, including electronic posters, a whiteboard movement and a water collection event. A donation drive for Ethiopian water projects was also organized. This paper aims to examine the effectiveness of RI Water Week as an advocacy event.

3. Methodology

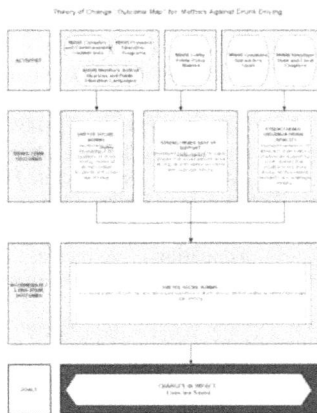

Figure 7: An example of Reisman's Outcome Map showing the processes of change

This paper will follow Reisman's Outcome Map (Figure 7) to examine the effectiveness of RI Water Week (Reisman, 2007). The desired outcomes of RI Water Week in the short-term and long-term will be established. The success of these outcomes will then be assessed through a series of metrics.

Quantitative metrics include the amount of donations collected and number of website views. Qualitative metrics include the quality of participant feedback (60 participants were surveyed).

4. Desired Outcomes

Short-term:

1. Increased attention on water issues (S1)
2. Significant donations to Ethiopian water projects (S2)
3. Increased awareness of water scarcity facts (S3)
4. Increased motivation to embark on water-saving (S4)

Long-term:

1. Sustained interest in water issues (L1)
2. Habitual practice of saving water (L2)

5. Short-term Outcomes Assessment

The activities of RI Water Week will be assessed against their short-term outcomes (S1-S4). The level of success is indicated as "very high", "high", " medium", "low" or "very low". Outcomes that are irrelevant to an activity are greyed out.

All choices made are explained with the use of quantitative and qualitative metrics as shown in Table 1.

Table 1: Short-term outcomes assessment of RI Water Week

S/n	Activity	Level of Success				Reasons
		S1	S2	S3	S4	
1	Thirst in Ethiopia Booth	High		Low		S1: The Thirst in Ethiopia Booth had 6 colourful cutouts to attract the attention of students. It was also placed in a high-traffic location (Canteen Walkway), where many students looked at the booth. 87% of students surveyed claimed that this booth increased their attention on water issues. S3: The Thirst in Ethiopia Booth consisted of statistics from World Vision Ethiopia. However, 65% of students surveyed claimed that the booth did not increase their knowledge of water scarcity. One student

#						
						commented that he "did not have time to read the long paragraphs fully".
2	H²OPE Donation Drive	High	Very High			S1: The H²OPE Donation Drive gave students a choice of which of 3 water projects to donate to. Letting them choose made them seek to understand and pay attention to water issues in Ethiopia. The blue ribbon souvenir also generated hype for advocacy. 89% of students surveyed claimed that this donation drive increased their attention on water issues. S2: After 6 days, $1072.60 in donations was collected. According to World Vision Singapore, this amount can pay the bulk of building a spring or sponsor hygiene lessons for a long period of time.
3	Save That Drop Movement	Low			Med	S1: Due to misallocation of manpower, few student-helpers took shifts to promote the Save That Drop Movement. By the end of the week, only 50 students had taken the pledge. 22% of students surveyed said that this movement increased their attention on water issues. S4: The Save That Drop Movement featured an interesting concept of spinning a wheel to make a pledge. For students who pledged, the wheel made them think about and commit to new ways to save water. One student said that the movement "inspired [her] to understand water issues more". However, due to poor student participation, this movement was unable to increase motivation to save-water significantly.

4	Water Issues Talk	Very High		Low	Med	S1: The talk by Mr Chew Men Leong, CEO of PUB, reached out to a very large group (1250 students). Mr Chew gave shocking details about Singaporeans' excessive water usage habits, which placed water issues under focus. 98% of students surveyed said that the talk raised their awareness of water issues. S3: The talk did bring up several statistics on global water scarcity. He also touched on the recent local dry spell. However, these facts were fairly general. 88% students felt that the talk introduced no few new facts on water scarcity. S4: By explaining the threats to local water security, Mr Chew conveyed the need for students to save water. One student mentioned, "Mr Chew's speech made us feel responsible to conserve water. However, it was draggy and thus less inspirational." 59% of students surveyed felt that Mr Chew's speech increased their motivation to save water.
5	Thirsty Tilaya App	High		Very High		S1: Thirsty Tilaya was able to generate hype about RI Water Week due to the 'fun' element. Based on game data, over 600 students played the game. 81% of students felt that the game increased their attention on water issues. An earlier release of the app could have improved outreach. S3: The fusion of water scarcity facts into gameplay made it fun for students to learn water facts. Also, these facts covered country-specific details that were new to students. 95% of students surveyed said that the game raised awareness of water scarcity facts.

6	Walk For Water	Very High	Low	Med	High	S1: From 14th April to 17th April, Walk For Water was publicized through a dramatic announcement and a self-filmed video. The video was looped in the canteen. Mass publicity for Walk For Water generated much focus on water issues. During the event itself, a jug of brackish water was purified for students to attract their interest. 91% of students surveyed said that Walk For Water increased their attention on water issues. S2: Only $60 in donations was collected from the Walk For Water. This amount was not able to pay a significant part of Ethiopia's water projects. S3: The Thirst in Ethiopia Exhibition was on display after the Walk. World Vision Singapore also set up a booth with information on water projects in Ethiopia. However, as most of the information given was general, and organized very wordily, students did not learn much about water scarcity. 11 out of 15 Walk participants said that the Walk increased their awareness of water scarcity facts. S4: After the Walk, participants were split into groups. Each group reflected on ways to reduce water scarcity. 12 of 15 Walk participants felt that they had a newfound responsibility to conserve water.

6. Predictions on Long-term Outcomes

In the long-term, RI Water Week should be successful in generating continued interest on water issues *(L1)*. 81% of students surveyed said that the campaign inspired them to learn more about water scarcity or water conservation over the next 12 months. As high levels of attention on water issues were created in the short-term, it is likely that the attention will grow into a genuine interest in the cause (Johnson et.al, 2005). Also, innovative activities such as Walk For Water made RI Water Week more memorable for students and such increases their long-term interest in water issues.

However, RI Water Week might be largely unsuccessful in encouraging habits of saving water *(L2)*. 87% of students surveyed said that the campaign did not inspire them to save water in the long-term. As the Save That Drop Movement reached out to less than 50 students, and the Water Issues Talk was viewed as "draggy", most students would not be motivated to adopt water-saving habits.

7. Limitations

Firstly, the survey was conducted on a sample size of 60 students. This may have been too small a number to reflect accurately the diverse views of the student body towards RI Water Week. Also, 80% of the sample comprised of Year 5 students, and Year 6 students only comprised 20%. This may also have reduced the accuracy of the representing students' views

Secondly, the research did not consider many secondary sources. Deductions were based mainly on metrics related directly to RI Water Week. As such, trends across advocacy projects were not examined in this research.

8. Conclusion

RI Water Week featured 6 activities, with varying levels of success in attaining desired outcomes. In the short term, the campaign was highly successful in increasing attention on water issues and raising funds for water scarcity. The placing of booths in crowded areas and the use of bright-coloured designs contributed to this success. The campaign was also successful in raising awareness of water scarcity facts, due to the use of technology. Thirsty Tilaya, a student-made app, allowed students to learn water scarcity facts while having fun.

However, the campaign had limited success in motivating students to save water in the short run. As many water saving tips were clichéd and Save That Drop Movement had limited manpower, few students were inspired to save water. The reflections on saving water by Walk For Water participants were meaningful, but it involved only 15 students.

Generally, students preferred innovative ideas (such as Thirsty Tilaya) over traditional outreach methods (such as the Water Issues Talk). The former was also more effective in raising awareness of water scarcity facts.

In terms of long-term desired outcomes, RI Water Week should be able to create sustained interest in water issues. However, it is very unlikely to instill water-saving habits in students due to its inability to motivate students to save water in the short-term.

References

1. Johnson, E. and Mappin, M. (2005). Environmental education and advocacy: changing perspectives of ecology and education. Cambridge: Cambridge University Press.
2. Reisman, J. (2007). A guide to measuring advocacy and policy. Unpublished manuscript, Annie E. Casey Foundation. Retrieved from http://www.aecf.org/upload/PublicationFiles/DA3622H 5000.pdf

Water defence in the Port of Rotterdam

Maud de Leeuw, Djimme van Etten, Anneloes de Brouwer and Lisa van Kruijl

St.-Odulphuslyceum, Tilburg, Netherlands, lisa@vankruijl.nl

Abstract

Our study focuses on 'hard and soft coastal management in relation to the Port of Rotterdam'. We also apply our findings to other major ports around the world. First we give a short introduction of the Port of Rotterdam and our nation's history in its battle with the sea. Then we discuss and analyze both ways of coastal management (which types are available, what are the advantages, disadvantages, costs etc.) and also compare them in order to find similarities and differences. Then we direct our attention to the Port of Rotterdam, Europe's largest port. This port is of high (economic) importance to the Netherlands and therefore it is vital to defend it against the sea. We examine and answer questions such as: which type of coastal management is best applied to the Port of Rotterdam: hard or soft? Or should we use a combination of both types? Finally we use our findings to look at several major ports all around the world.

Introduction

This means the purpose of this study is to examine what is the best way to defend the Port of Rotterdam. After finishing our investigation, we want to be able to answer our main question:

"What is the best way to defend the Port of Rotterdam against the sea?"

Method

For this study, we have chosen to do some literary research and also to interview two people, Prof. Bas Jonkman, who is a professor at the Technical University of Delft and is an expert in everything that has to do with water engineering, and Tjitte Nauta, who works at Deltares, a big company in the Netherlands that takes care of the water defence in the Netherlands.

Investigation and Discussion

1 The Dutch and their fight against water

The Dutch are known for their advanced coastal management techniques. Even though half of the country is below sea level, there has not been a flood in the past 60 years. This is a direct result of the dikes, dams, flood barriers and other water defences which can be found throughout the country. However, all these mega structures did not just appear by themselves. In the following part, we describe how coastal management got an increasing influence on the Dutch culture and landscape.

1.1 History of the Dutch coastal management

The Dutch government has been trying to protect their coastal areas against the North Sea. Due to the fact that a major part of the Netherlands lies below sea level, the country is extremely vulnerable. In fact: without any form of coastal protection, at least half of the country would be permanently flooded. Some of the most important cities of The Netherlands such as Amsterdam, The Hague and Rotterdam, would be destroyed by the water. A big problem is the Dutch soil constantly experiencing soil subsidence due to activities such as gas and ground water extraction but also due to the position of the Netherlands in Europe. While the Alps in France and Austria gradually rise, the Dutch land comes to lie increasingly lower. The Netherlands is often called "the drain of Europe" due to this process.

The Dutch have been applying coastal protection for centuries (Wikimedia Foundation, 2014). The first forms of coastal management were probably already used around the year 1000. The Dutch built dikes alongside the big rivers to make sure they would not flood. However, not many people knew much about building dikes at that time. This resulted in badly built dikes. For a long time, the dikes were adjusted to previous floods. If a flood occurred, the dike would be raised, often with half a meter.

The Netherlands has witnessed terrible floods. To name some: in 838, the North-Western parts of the provinces flooded and over 2.400 people were killed. In 1212, the province of North-Holland flooded. This is known to be one of the most terrible floods ever in the Netherlands: a frightful amount of 60.000 people died. More recently, in 1877 the northern province of Groningen flooded and 51 people died. The last flood that occurred in the Netherlands was one of the greatest disasters that had ever happened; the North Sea flood of 1953 in which 1836 People lost their lives.

During following centuries, the quality of the dikes substantially increased. The Church was also partly responsible for the increasing quality of the dikes. The church possessed large amounts of money. They could hire good technicians and architects who could build proper dikes. The Church could also bring together many people. They all worked together to build dikes that would actually protect the coastal areas. In around 1250, most of the Dutch dikes were of proper quality and could easily weather storm surges.

Modern dikes are made out of concrete and stone. However, once, most dikes were made out of clay and wood. Because of the wood, these dikes were known as hard coastal management, which we will discuss later on.

At the beginning of the 18th century, most parts of the Netherlands decided to not use wood anymore. The reason was that the Dutch VOC, the largest trading company in the world at that time, brought naval ship worms along from their trips to indigenous destination. It spread in the Netherlands and damaged wooden dikes severely, many dikes became instable and/or collapsed. The wood was replaced by big blocks of stone.

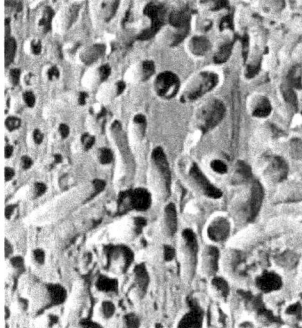

Figure 1: Naval ship worms

Around the same time, 'polders' were built in the Netherlands, another technical miracle. A polder is an area that is completely surrounded by dikes and sluices. Using these dikes and sluices, the water level can be determined, by opening a sluice, water will leave the area so that the water level will lower. ''Polder mills'' were used for the same purpose, the water level could be determined because the mills were able to pump water from one place to another. One of the most famous polder mills in the entire world are the mills found at Kinderdijk, which have been a UNESCO World Heritage since 1997.

In 1912 the research department of rivers, estuaries and coasts was created by Rijkswaterstaat (Ministry of infrastructure and environment). However, this was mainly done with shipping in mind, not necessarily for coastal defences. In the years 1906 and 1916, the Dutch coast was struck by a flood. Even though the dikes were much stronger than they had been before, the wood had been replaced by basalt, they were still not high enough. It was decided the dikes should be raised, the only problem was the money which would have to be invested. A relatively cheap and effective solution had to be found. Sir Muralt, head of the technical water department Schouwen between 1903 and 1913, came up with a solution: little walls made out of 3 or 4 layers (approximately 1 meter) of concrete. A great benefit was the fact that the dikes now did not have to be broadened. These little walls were named after their inventor; "Muraltmuurtjes". Approximately 120 kilometres of sea dikes have been raised by Muralt walls between 1906 and 1935.They were quite effective, but in many places this was the only real change made to the sea dikes until after the flood of 1953. The lack of focus on the coastal defences after the building of the Muralt walls was mainly caused by the fact that the Dutch Ministry of infrastructure and environment was occupied with the reclamation of the Zuiderzee. After the high water levels in 1943, when the sea reached over large parts of the Dutch dikes, something was bound to happen, but no real action was taken. Of course, the second World War also played its part in the delay of carrying out eventual plans.

Figure 2: Dike with Muralt wall

1.2   The flood of 1953

The flood of 1953 is an important part of Dutch history and necessary in order to fully understand in what way the present coastal defences have been established.

On the 30th of January of that year arose a depression (a low pressure area) around Iceland. This depression gradually travelled south and during that night it already contained a large storm field. Soon the entire North Sea was surrounded by this depression. On the 31st of January a hurricane approached the northern and eastern coast of Scotland and this hurricane moved towards the Netherlands. At this time it happened to be spring tide. This is a process in which the water levels rise due to the gravity caused by the moon and the sun. The gravity is at its maximum because the moon and the sun are in line. This spring tide was enhanced by the storm that derived from the hurricane that approached. During the evening of the 31st of January wind force 10/11 was measured along the entire Dutch shore and water was already crossing the levees at some places. Warnings were being sent out, but many municipalities did not receive these warning. None of the weather stations were active during the night and most of the smaller stations were only active during daytime. Many people did worry, but most of them thought the dikes would be high enough, even though the water now crossed the dikes during low tide. At night the water levels measured 4.55 meters above sea level at some places. The water was raised around 3.10 metres by the wind alone. The dikes were certainly not capable of holding back such great force. Around 3 to 4 am the first dikes collapsed. This mostly happened in the following places: Oosterschelde, Grevelingen and Hollands Diep. The water swiftly flowed into the polders. In some polders the water reached a height of 2 to 4 metres. In total, 89 dikes collapsed, spread over 187 kilometres of coastal area. Even though it was not a densely populated area, 865 people died in the province of 'Zeeland'.

The North Sea flood of 1953 has had a great impact socially, economically and geologically, but also in the minds of the Dutch: "this never again!" This disaster could even have been even worse. The tide was not at its maximum height the moment the disaster occurred, but still the dikes collapsed in large numbers.

1.3   The establishment of the Delta works

After the North Sea flood of 1953 it became apparent something important had to be done, the coastal defences had to be changed for the better. A disaster like the flood of 1953 should never be allowed to happen again.

The Delta plan was already created before 1953, but the execution had been delayed. Only after the disaster actions were taken. The Delta commission was appointed by the Minister of Infrastructure and Environment. This commission had to advise the minister about which parts of the Delta plan had to be executed. In 1960 the commission advised to establish standards which resulted in the establishment of flood risks, which means a flood once per x years and also takes the consequences into account, 1 flood in 125.000 years is now seen as the optimum level of protection. According to the commission 1 in 10.000 years would be sufficient. This norm is still valid today. The official written part was now finally completed and the actual execution of the Delta plan could now be initiated: the construction of the Delta Works. In 1986, the world-famous Delta Works were completed. This defense mechanism, consisting of flood surge barriers and dikes, protects almost the entire southern part of the country. These Delta works were captured in the Delta law that was founded on the 8th of May 1958.

This is an overview of the in the Southwest-Netherlands realized Delta works:

1. Waterkering Hollandse IJssel
    (1954-1958)

2. Zandkreekdam
    (1957-1960)

3. Veersegatdam
    (1958-1961)

4. Grevelingendam
    (1958-1965)

5. Zeelandbrug
    (1963-1965)

6. Volkerakwerken (Hellegatsdam met
Hellegatsplein, Volkerakdam, Haringvlietbrug)
    (1955-1977)

7. Haringvlietdam
    (1956-1972)

8. Brouwersdam
    (1963-1972)

9. Stormvloedkering Oosterschelde
    (1967-1986)

10. Markiezaatskade
    (1980-1983)

11. Oesterdam
    (1977-1988)

12. Bathse Spuisluis
    (1980-1987)

13. Philipsdam
    (1976-1987)

14. Maeslantkering
    (1991-1997)

15. Hartelkering
    (1993-1996)

16. Westerscheldetunnel
    (1998-2003)

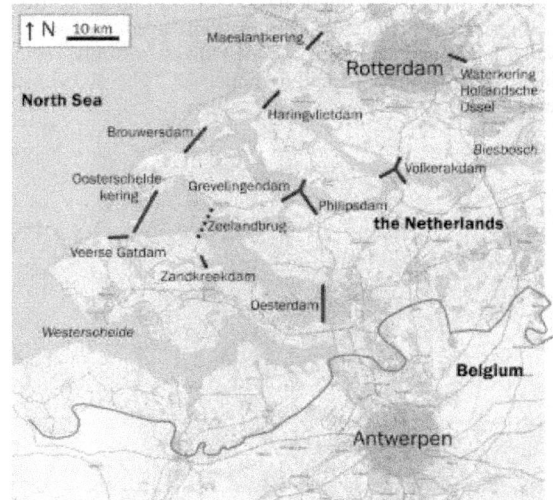

Figure 3: The Deltaworks

1.4 Water management as an export product

The Dutch knowledge regarding water management is recognized and appreciated all over the world. Due to this knowledge Dutch companies focusing on water management can be found worldwide. For example, after the destructive hurricane Katrina hit the city of New Orleans in 2005, multiple Dutch water-managing agencies were asked for help to rebuilt and strengthen the dikes.

There are a lot of Dutch companies working on projects all around the world. For example, in 2011 the Dutch company 'Deltares' worked on building and strengthening dikes in the city of Jakarta, the main capital of Indonesia. Because of the work of Deltares, a city with almost 10 million residents will be protected against the destructive sea.

In the year 2000, a Russian submarine, the Kursk, sank after an explosion and 118 people died. A Dutch team was asked to recover the bodies and the wreck of the submarine. A more recent example (January 2012) is when the Costa Concordia, a cruise ship, partially sank near the coast of Tuscany in Italy. A Dutch team, sent by the company Smit International, was asked to come to Tuscany. This team pumped 2,380 tonnes of heavy fuel oil out of the ship. These examples are proof that the Dutch do not only have knowledge about coastal management, but are also proficient in everything that has to do with the maritime sector.

The Dutch also see the economic importance of water management. The Netherlands has connections with numerous countries partly because of its extended knowledge regarding water management. This is also the case in Singapore where some Dutch companies can be found. For example, the leading Dutch applied research institute, Deltares and the National university of Singapore have created a knowledge alliance known as NUSDeltares, which focuses on the most pressing water and soil issues confronting the world today.

The king of the Netherlands, Willem-Alexander van Oranje-Nassau, is seriously interested in the Dutch water culture. He is specialised in water resource management and has been chairman of the Advisory Committee on Water to the Dutch Minister of Infrastructure and the Environment for eleven years and chairman of the Secretary-General of the United Nations' Advisory Board on Water and Sanitation for seven years. Often, when he is on a state visit, he will also visit some local water-concerning companies with other Dutch businessmen. During these state visits, sometimes new deals are made between Dutch and local companies. Water is essential for the Dutch. The Netherlands and water: a love-hate relationship.

2   Hard and soft coastal management

Naturally, there is something called a 'dynamic balance' in nature: sometimes the sea takes a piece of land, and sometimes it gives back a piece of land. Despite this 'balance', it is still important to defend a densely populated area against the obtrusive sea, especially when it is located near the coast. To protect the Dutch coastline, there is a choice between roughly two methods of coastal management: hard and soft coastal management. Hard coastal management strategies are for instance building dikes, breakwaters and groynes, while soft coastal management strategies usually involve beach nourishment or protection by plants at vulnerable places. Both ways of coastal management have an influence on the existing environment and can sometimes even disturb the ecosystem.

Because of the adjustments made by man, the sea is forced to find a new balance. That is why it is important to investigate the impact the different ways of coastal management will have on the sea's balance before the coastal management is actually built. Science is already rather experienced in predicting 'coastal dynamics'. By giving proper attention to aesthetic and ecological aspects, a coastal management mechanism can be applied in a decent way. We should also not lose sight of the touristic and economic aspects. However, the question that remains is what hard and soft coastal management actually entails and what the advantages and disadvantages might be.

2.1   Hard coastal management

The innovative and impressive ways the Dutch apply hard coastal management are known all around the world. For example the Delta Works, containing dikes and flood barriers protecting the provinces of Zeeland, North-Brabant and South-Holland against the sea are known all around the world and have even been listed as one of the ''Seven Wonders of the World'' by the American Society of Civil Engineers.

Superficially, 'hard coastal management' contains all hard objects used to protect the Netherlands against the North Sea. These are, among others, dikes, dams, flood barriers, groynes and breakwaters and they can all be found in The Netherlands. However, they did not simply appear out of nowhere, a long history preceded these inventions. In the next part, it will become clear how hard coastal management has an increasing impact on the Dutch culture and landscape.

Figure 4: The Oosterscheldekering, a part of the Delta Works.

2.1.1   Types of hard coastal management

As already described, in the Netherlands many types of hard coastal management (Wikimedia Foundation, 2014) can be found. Following, all these types will be described.

- Dikes: a dike is a flood barrier that is built along the shore or along a river and it always runs placed parallel to a 'water'. The main part of a dike is stone and/or concrete. Sometimes, a layer of asphalt is placed upon the dike so that transport across the dike is possible. Dikes are supposed to protect the land that is located behind the dike from the water. The Dutch are always working on their dikes. At the moment there are a lot of repair projects ongoing. In 2015, a total of 325 kilometres (= 201 miles) of dikes will have to be repaired; the Dutch coastline has a total length of around 357 kilometres.

Figure 5: A dike with a road on it. On the left the Markermeer. On the right the Oostvaardersplassen.

- Dams: a dam is an object that crosses a 'water', like a river. With a dam, the water level on both sides of the dam can be determined, for example by opening or closing a sluice. When there is a storm surge, the dam and its sluice will be closed so that the water cannot get to the inhabited area. One of the most famous dams in the Netherlands and in the world is the Oosterscheldekering, a part of the well-known Delta Works.

- Storm surge barriers: a storm surge barrier is a construction that is specifically aimed to protect a (populated) area during storm surge and high tide. During such a situation, an extreme amount of water is stowed towards the land. Storm surge barriers are specifically made to resist these amounts of water. (Parts of) the Delta Works are the world's most famous and one of the best storm surge barriers. Storm surge barriers are certain types of dikes and dams, but not all dikes and dams are storm barriers.

- Groynes: the types of hard coastal management described before are all ways to directly protect the land from the water. However, groynes indirectly protect the land from the water. Still, they are very important when talking about coastal protection. Groynes are narrow, straight constructions made from stones that are placed right-angled on the shore. They are often seen at beaches (Vliz, 2004-2014).

A groyne deflects heavy sea currents from the shore. Because there are no heavy currents near the shore, less sand from beaches and dunes (a form of soft coastal management) will be flushed away. If this were to actually happen, dunes could become instable and even collapse and that would significantly increase chances of flooding. Groynes are a good example of a collaboration between hard and soft coastal management.

Figure 6: Groynes in the Dutch river the Rhine.

- Breakwaters: a breakwater is a construction made out of stone that mainly protects harbour areas, like the Rotterdam harbour. A breakwater can be compared with a stone wall that is built around the entrance of the harbour. In the breakwaters, some openings are made so that the ships can still enter the harbour. When a strong sea current or a high wave hits a breakwater, they are slowed down and some will be rebound so that the water inside the borders of the breakwater will remain calm. As a result, the chance that the harbour area will flood, is very low.

2.1.2. Advantages and disadvantages

Hard coastal management has some advantages. A first advantage is that the hard constructions, such as dikes and flood barriers, are very trustworthy because they are made with solid materials such as stone and concrete. These materials will last for years and they can also resist heavy sea currents, storm surges etc. very well. Another advantage of hard coastal management is that these techniques will ensure that sand replenishments (more about this at 'Soft coastal management') will be retained for longer. Groynes for example cause strong sea currents to stay away from the coast, so that the sand replenishments will not be flushed away.

Figure 7: In 1421, the dikes collapsed. The St. Elisabeth Flood flooded the city of Dordrecht.

There are of course also some disadvantages about hard coastal management. Firstly, hard coastal management techniques can affect the environment in which they are placed in a negative way. A huge concrete structure like a dike simply does not belong in a natural environment and it is not hard to imagine that this will have negative implications. For example, a dam in a river hinders a free current of the water and it makes it impossible for fish to get to the river that is on the other side of the dam. This can be very bad, because there might be large supplies of food on the other side of the dam. If the fish cannot get there anymore, they will die or their lives will at least become much harder.

Another disadvantage of hard coastal management, is that it will disturb the continuity and flexibility of the area in which it is placed. The coast line is always ''moving''; sand is eroded at one place and is brought to another place, vegetation disappears etc. This will make sure that the appearance of the coast is always changing, but the hard coastal management techniques are not changing! They will always have the same shape and will always stand on the exact same place. When the coast changes, but the management techniques do not, this can have negative results; the dikes, dams etc. are not well-connected to the newly-formed coast which can decrease their efficacy and efficiency.

The Dutch approach of (hard) coastal management seems to be very successful. Since the Dutch really focussed on their shore protection, after the North Sea flood of 1953 which killed 1,836 citizens, not a single flood with fatalities has happened.

2.2 Soft coastal management

Apart from hard coastal management, there is also soft coastal management. Due to the rise in sea level coastal erosion is a natural process in the Netherlands. Since the first Coastal Policy (1990), the deterioration of the coastline is combated with 'soft' measures through beach nourishment. This coastal defence policy is based on the dynamic maintenance of the basic coastline of the Netherlands ('BKL'). Since 2001, there is a policy that the coast naturally grows along with the sea level rising. Only when there is no other solution, the coast is defined by hard constructions (hard coastal management). This was decided because hard coastal management is very costly and therefore leads to high maintenance costs.

2.2.1 Development of soft coastal management

For hundreds of years the dunes and polders have protected the coast against the seawater. Nevertheless, people had often lent Nature a helping hand to protect the country from the sea. After a storm, Nature herself was not able to recover quickly enough to continue the protection and one worked very hard to restore the dunes. In spring, for example, the dunes were planted with marram grass, to make sure that the dunes were ready to protect the country in autumn and spring, when the storm season began.

Figure 8: Marram grass on a dune.

Because of the increasing technical possibilities to fight against the sea, people increasingly choose to fight 'hard against hard'. (Baptist, 2012) The stable boundary between the sea and the land was statically realized by the use of hard coasts. For quite a while now we realize that a good coastal defence not only can be achieved with hard coastal defence but also with a dynamic equilibrium. In this case the sea is no longer seen as an enemy but the sea plays an active role in the defence process by emphasizing the interaction between the sea, the beach and the dunes.

One of the most important and environmentally friendly forms of soft coastal defence is beach nourishment (Wikimedia Foundation, 2014), a process by which sediment (usually sand) that is lost because of long shore drift or erosion is replaced with sources (sand) from the sea. Trenches that lead to harbours are important for the winning of sea sand. The trenches have to be dredged continuously to make sure that they remain deep enough to let ships go through them. Instead of dumping the dredged sand somewhere in the ocean without using it, it is now useful in sand replenishment.

Huge ships with a kind of vacuum cleaner suck the sand from the sea bottom. This sand is transported to the coast in large pipes and then it is sprayed on the beach. Previously, only damaged places were repaired but nowadays an extra layer of sand is applied, which serves as a wear layer. This layer captures the erosion for a long time which results in the dunes retaining their size.

Beach nourishment is very efficient, especially when it is applied in combination with hard coastal defence. It can be used anywhere on the beach. However, a good look will have to be taken at the circumstances of the beach, because not everywhere on the beach sand replenishment is recommended or cheap. In some cases, the replenished sand is dissipated by the water very quickly. In these places, beach nourishment is needed repeatedly to keep the coast safe which is not financially viable.

Figure 9: Beach nourishment

Research is being done at this moment to improve beach nourishment. Scientists are examining whether it is possible to improve the quality or composition of sand. They have discovered that adding small carbon elements to the sand ensures that the sand is more resistant to water. By making sand heavier, it is more resistant to erosion and finally it can also help to add micro-organisms to the water to ensure that the sand is consolidated in an organic way.

Letting overgrow the dunes with vegetation (to make sure that the sand is retained better) and the different types of beach nourishment are soft coastal management. Especially beach nourishment appears to be the most effective and stable form of soft coastal defence. This form is used the most in the Netherlands and is therefore the most important form of soft coastal defence.

2.2.2 The current approach, a success?

In the research report 'De levende natuur (2011)' (the living nature), by M.J. Baptist and W.A. Wiersinga, students at Wageningen University, it has been proven that the current types of soft coastal management by beach nourishment are successful regarding safety and that it fits well in the natural coastal dynamics. On top of that, soft coastal management is flexible, whereby it is possible that through monitoring the state of the sea near the coast, one can correct and drive the coastal defense. This is also called adaptive management. Adaptive management makes it easier to take care of the coast and it also ensures that the protective factor of the soft coastal management is not weakened.

The current size of beach nourishment (12 million cubic meters of sand a year) is based on the minimal use of nourishments; they will only be used when the coast is not able to defend itself in a natural way. For the other parts of the Dutch coast, natural processes are preferred. However, keeping the expected sea level rise in mind, this size of beach nourishment is not enough to maintain the basic coastline of the Netherlands. As a result of the above mentioned, the coast will have evolved into a craggy coast by 2040. There will be dune erosion, dune breakthroughs and slufters. A slufter is an inland area with salt sea water. This sea water reached the slufter via holes in the dunes. Because the morphology of the shallow sea coast develops whimsically, there will be a lot of different environments and a variety of species.

As the basic coastline cannot be maintained with the current size of beach nourishment, there is also a study that looks at what would happen to the Dutch coast if the size of beach replenishment were increased. This appears to be ecologically possible as well. However, when volumes greater than 50 million cubic meters a year are used this leads to a repeat period (the time between two replenishments) that is shorter than the biological recovery duration of life at the bottom of the sea. This ensures that sand replenishments of this magnitude are threatening the ecosystems of the shallow sea coast. Mega nourishment may also be a solution to maintain the basic coastline, but this also has a limit. This limit is determined by the morphological recovery period of coastal waters. When there are too many mega nourishments, the biodiversity of the animals that live in the water near the beach will suffer. The mega replenishments and replenishments greater than a hundred cubic meters a year cannot be performed without major consequences to the coastal sea.

The Dutch government has therefore decided not to use coastal expansion on a large scale. As the size of the sand replenishment cannot be increased without major consequences – and the basic coastline cannot be maintained with the current size of the sand replenishment – it is very hard to maintain the basic coastline only with beach nourishment.

2.2.3 Advantages and disadvantages

There are some advantages to soft coastal management. (Beerda, 2012) A first advantage is that soft coastal management is a good option to use in areas where there is a lot of tourism. By using beach nourishments the recreational beach is widened and as a result there will be more space for tourism on the beach. Another advantage is that as long as the replenished sand remains in its place the coastal defence is effective. There is no damage caused to the structures behind the beach.

However, there are also disadvantages when using soft coastal management. Hard coastal management has a good resistance against the weather but soft coastal management is less resistant. The replenished sand can erode because of storms and wave impact. Moreover, beach nourishment is rather expensive and must be repeated every now and then. Besides that, the composition of the sand that is used for the replenishment is often different from the composition of the 'normal' sand on the beach which makes it hard to find sand that fits the beach.

2.3 Costs of hard and soft coastal management

It is very difficult to say what the specific costs of coastal management are because there is no specific information. However, it is obvious that the construction and maintenance of all structures that have been described are very costly. The Netherlands will never really finish improving its coastal defence. In a 2010 report published by the Dutch government it was stated that the costs for the entire Dutch 'water systems' in 2010 were an incredible 7 billion euros. It is expected that these costs will continue to rise (Rijkswaterstaat, 2010).

3   The Port of Rotterdam

The Netherlands is divided into twelve provinces. One of them is Zuid-Holland (South-Holland) and this is the province where Rotterdam is located. The western side of the province lies on the North Sea.

The Port of Rotterdam covers a total of 10.570 hectare and has a length of 40 kilometres, which makes it the biggest port of Europe and of course the Port of Rotterdam is also of great economic value. It consists of various harbour basins and business parks. The Port of Rotterdam provides the opportunity to other nearby industries to supply and export goods. Moreover, goods are often stored in the port itself and then later transported.

The Port of Rotterdam is situated on the mouth of the Rhine. This river connects Rotterdam with a hinterland where over 460 million people live. A lot of the goods arriving at the port are transported to the Ruhr district in Germany and to Antwerp in Belgium.

Figure 10: Figure 10 - the Dutch country with the red dot indicating Rotterdam.

The Port of Rotterdam consists of the following parts:

- Port in the old town (until 1872)
The Hoogstraat runs right across Rotterdam. This street separated Rotterdam into a 'land city' and a 'water city'. In the land city, people lived. In the water city on the other hand, harbours were constructed. During the 18th century various harbours were constructed and the amount of ships visiting Rotterdam grew by year. Due to the industrial revolution (second half 19th century) increasingly bigger and bigger ships were built. There was a need for moorings. Between 1847 and 1854, the Willemskade and Westerkade were built and also some new harbours were under construction.

- De Nieuwe Waterweg (1872 – present)
Before de Nieuwe Waterweg (the New Waterway) was constructed, ships that had to travel from the North Sea had to make a detour in order to arrive at the Port of Rotterdam. This sometimes brought along many hardships because

rivers were often bogged down. In the year 1858 mthe plan to dig through the dunes of Hoek van Holland in order to create the New Waterway was brought up by Pieter Valand. This Waterway was 4,3 kilometres long. The New Waterway was fully completed in 1889. The Port of Rotterdam now had a direct connection to the North Sea.

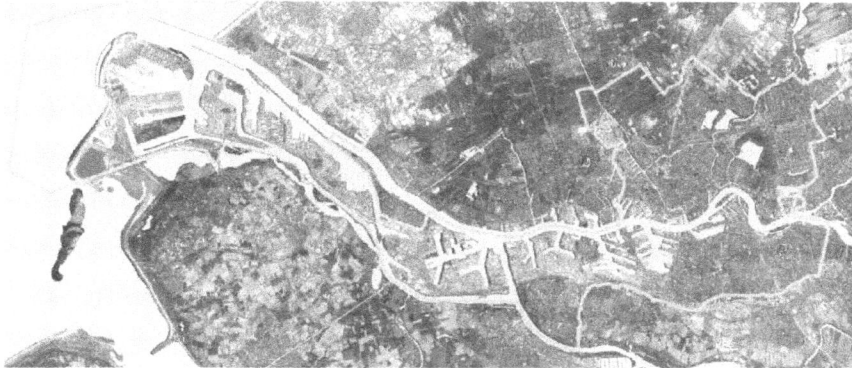
Figure 11: Map of the Port of Rotterdam

- Construction of the ports

After the construction of the New Waterway, new ports had to be constructed otherwise, the boats would not have a place to transport the goods. The isle IJsselmonde was the first place where moorings and terrains for goods were built. The first oil port was constructed in 1929 in Pernis. This port was important for the petrochemical industry. In 1967 the Europe Container Terminal, which is the biggest container company of Europe to date, was opened. This led to the Port of Rotterdam becoming more well-known all over the world.

Freight traffic became more and more important after the Second World War and as the harbours of IJsselmonde did not quite cover the amount of space needed for the extra traffic, new ports were created in the Botlek-area around 1947. All of these were cargo ports, with the emphasis on the petrochemical industry.

Figure 12: Map of IJsselmonde

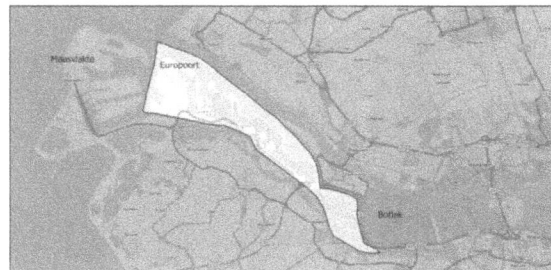
Figure 13: Map of the Botlek-area

- Europoort (1957 – 1968)

The sea ships kept growing in size, the waterway around Botlek became too small for them and they could no longer pass it. This led to Rotterdam constructing the Europoort area. This area is focused exclusively on processing crude oil and crude oil trans-shipment. In the Western part of the Europoort area some land has been designated for the storage of grain and ores.

- Europoort (1957 – 1968)

The sea ships kept growing in size, the waterway around Botlek became too small for them and they could no longer pass it. This led to Rotterdam constructing the Europoort area. This area is focused exclusively on processing crude oil and crude oil trans-shipment. In the Western part of the Europoort area some land has been designated for the storage of grain and ores.

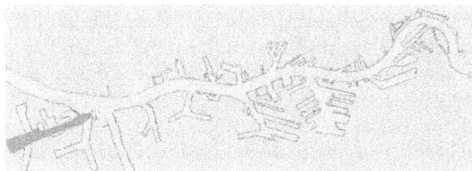
Figure 14: The Europoort area

- Maasvlakte I (1965 – 2008)

The most recent expansion of the Port of Rotterdam took place during the 60s of the past century. It involves the construction of the Maasvlakte. This area is located next to the mouth of the river the Maas and lies on the North Sea. Maasvlakte 1 occupies a surface of 3000 hectare and nowadays four container terminals are located here.

- Maasvlakte II

The current area of the Port of Rotterdam does not offer enough space anymore to accomodate new companies and already existent customers who want their company to grow. Especially for the sake of the deep sea related container sector, the chemical industry and the distribution parks space is needed. Maasvlakte 2 will be built deeper in the sea, which is more convenient for larger ships. More information about Maasvlakte 2 can be found in the next section.

All together it can be said, that the Port of Rotterdam has been expanding from the small docks in the old town (mid-19th century) to and even into the North Sea (present), with the Maasvlakte and its huge terminals.

Thus accommodating the changing demands of ever larger ships and wider and deeper shipping lanes.

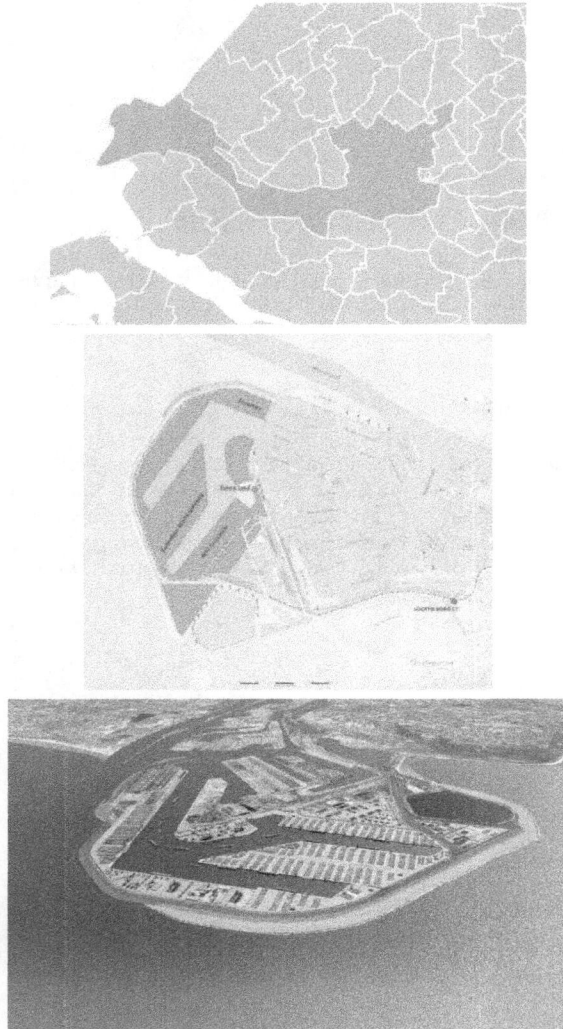

Figure 15: Maasvlakte I and II

4 Types of coastal defense applied in the Port of Rotterdam

As mentioned in the previous section, the Port of Rotterdam covers a gigantic area of 10.000 hectares. Because of this hundreds of ships can and do arrive at the port on a daily basis. To be exact: in 2013 an astronomical amount of 440,5 million tons of products passed the Port of Rotterdam, making it Europe's biggest port by far. The second biggest port in Europe is the Port of Antwerp, where ''only'' 190,8 million tons of products were treated. It is not hard to understand that the Port of Rotterdam is of high economic importance to the Netherlands. Therefore, it is

very important that the Port of Rotterdam is a very secure and safe place. A lot of types of coastal defence are applied in the Port of Rotterdam. The coastal defence systems mainly consist of storm surge barriers and dikes. In this section, every type of coastal defence applied in the Port of Rotterdam, will be discussed.

4.1 The Europoortkering

The Europoortkering is part of the Deltaworks. The Deltaworks is a plan that was created after the North Sea flood in the Netherlands in 1953, as mentioned in the previous section. It mostly consists of barriers which block the sea from connecting to large parts of the mainland and thereby reducing the Dutch coastal line. The construction of the barrier started in 1991 and ended six years later (Rijkswaterstaat, 2014).

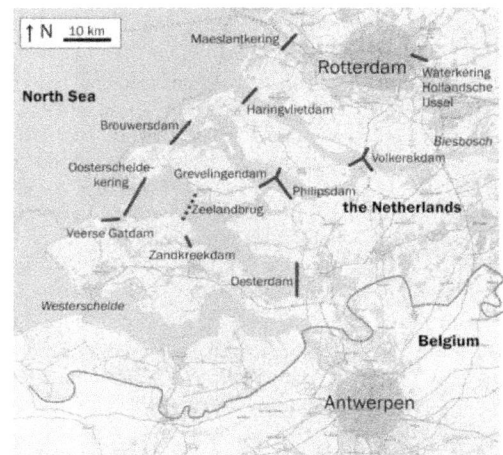

Figure 16: The Deltaworks. The blue lines indicate several parts of the Deltaplan, which consist of dams, sluices, locks, dikes, levees, and storm surge barriers.

4.2 The Maeslantkering

At the moment, a storm surge barrier protects Rotterdam against high sea levels. This storm surge barrier is named "the Maeslantkering". The Maeslantkering is part of the Europoortkering which was made in order to protect great parts of 'South-Holland'. The Maeslantkering storm surge barrier is located in "de Nieuwe Waterweg" (the New Waterway), which is the route from the North Sea to the city of Rotterdam. You can see where the Maeslantkering is located on the map on the right.

When the Maeslantkering was built there were, of course, already dikes present which had been built along the length of the new waterway. However, they were found not to be high enough in order to comply with the standards. Raising them more would be very expensive, because raising dikes means broadening them as well, as the centre of gravity has to remain low. A barrier would have to be built.

Because of its function, the new waterway could not simply be closed. The Port of Rotterdam had to stay easily accessible, especially since it was the largest port in the world at that time (1991). So the Maeslantkering would have to be more flexible than other barriers that are part of the Deltaworks such as the "Oosterscheldekering" had been. That is why the Maeslantkering now consists of two huge revolving doors which are able to entirely close the New Waterway, in case the water rises to a dangerous level. When the Maeslantkering is closed it can resist a tidal wave which is 5 metres above the average sea level (Rijkswaterstaat, 2014). Every year, the Maeslantkering is tested. In 2001, researchers found that there was a chance of 10% that the storm surge barrier would not close properly. When the barrier does not close during a tidal wave, many lives can be at risk. From then on, many improvements have been made; the control of the surge has been improved and the Maeslantkering was checked and repaired more often. The chance that the Maeslantkering does not close when it needs to, is only 1% right now. To keep the risk this low, every year a test takes place to check whether the storm surge barrier still works as it should do. The Dutch government has stated that because of all the improvements and safety measures Rotterdam will flood only once in 7000 years. So we can say that the Maeslantkering is extremely safe and important for the maintenance of the Port and the city of Rotterdam.

Figure 18: The location of some dams and storm surge barriers. The Maeslantkering and Hartelkering are storm surge barriers.

Figure 19: The Hartelkering consists of two white plates (see image), with a width of 49,3 en 98 meters (= 164 and 321 feet). When there is a high sea level, for example during a storm, these plates move down so that the water is blocked. The Hartelkering can resist a tidal wave that is 3 metres above average. When this barrier is not closed the plates are 14 metres above the water level.

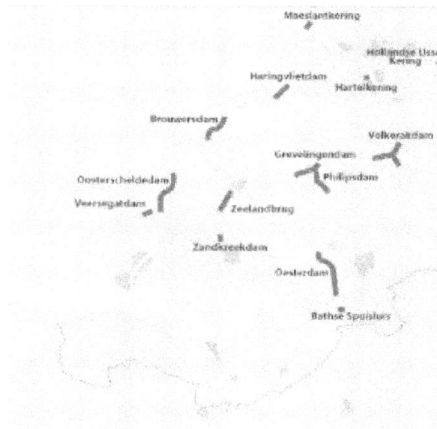

Figure 17: The Maeslantkering.

4.3   The Hartelkering

When the Maeslantkering was being built, at the same time another storm surge barrier was under construction: the Hartelkering. The location of this barrier can be seen on the map on the right. It is located more landward than the Maeslantkering. The Hartelkering has a very important function. When the Maeslantkering is closed during a heavy storm, more water will flow landward towards the Hartelkering. If the Hartelkering were not there, all these giant amounts of water would flow immediately to Rotterdam and other (populated) areas in South-Holland. These areas would flood and it is very plausible that many people would die. Because of the Hartelkering, the water will be stopped and the safety of the people in Rotterdam and South-Holland is ensured.

4.4   Dikes and dams

The two storm surge barriers are not the only buildings that protect (the Port of) Rotterdam from the dangerous sea. There are many other constructions that also take part in the protection, for example dams. On the map see the location of many dams that protect Rotterdam and the other parts of South-Holland can be seen. Furthermore, there are also a lot of dikes that protect the city of Rotterdam and surrounding areas. There are dikes at Rozenburg that protect the village of Rozenburg and some parts of the port. These dikes were strengthened at the same time the Maesland- en Hartelkering were built. Also the Brielse Maasdike, next to the Hartelkering, was strengthened. It was made 3 meters (= 9,8 feet) higher. These dikes and dams do not protect the port, because the port is located outside the dikes to make sure that all the ships and boats can easily reach the Port.

163

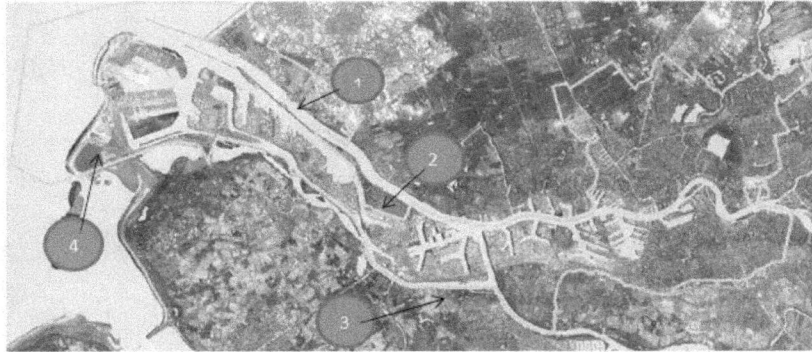

Figure 20: A satellite photo of the Port of Rotterdam.

1 = The Maeslantkering.　2 = Rozenburg.　3 = The Hartelkering and the Brielse Maasdike.

4 = Maasvlakte I (from 1996 to 2013, more land was ''created'' over here: Maasvlakte II).

4.5　Maasvlakte II

In 1996 it was already stated by the Dutch government that the Rotterdam harbour would outgrow its borders in the near future. Therefore, more space had to be created, Maasvlakte 2 was the solution. Maasvlakte 2 is a piece of land which would be created by reclaiming land from the North Sea. It was created with a certain figure in mind: Maasvlakte 2 had to be innovative, feasible and economically smart. Maasvlakte 2 would cover about 2.000 ha.

1.000 ha would consist of infrastructure, such as seawalls, water- and railways, roads and port basins. The other 1.000 ha would be used in order to create space for industrial sites. On May 22, 2013, the Second Maasvlakte was officially inaugurated.

Figure 21: The former queen of the Netherlands, HRH Princess Beatrix, inaugurated Maasvlakte II in 2013.

The defence of Maasvlakte 2 consists of two parts: the western and northern sea defence. Since with the building of Maasvlakte 2 beaches are lost, this loss has to be compensated for. Therefore, the Western sea defence involves the construction of an artificial dune. This type of sea defence is actually not preferred for the northern sea defence, because it is harder to calculate what will happen to the dike in maximum storms, or so called design storms.

Therefore, it is not certain how much time and money will have to be invested in maintenance operations.

Maasvlakte 2 is protected by a partly soft and partly hard seawall (Havenbedrijf Rotterdam N.V., 2014). There are beaches and dunes that form the soft part, but also pebbles, stone, quarry stone and concrete blocks that form the hard part. A special and unique form of coastal management has been applied at Maasvlakte 2. It is a mixture of hard and soft coastal management. In the picture below can be seen what the coast and the coastal defense at Maasvlakte 2 look like.

On the right, you can see a part of Maasvlakte 2. It is situated at +5,5 metres (= 18 feet) above the average sea level. Situated left to Maasvlakte 2, you can see an increase (at +13,7 metres = 45 feet) and then a sloping shore. This is mainly made up of sand, see nr. 1. On top of the sand there are layers of gravel, see nr. 2, and boulders, see nr. 3.

On top of the layer of boulders there is a layer of stones, see nr. 4 and 6. The stones in number 6 ensure that the large pile of concrete blocks, nr. 5, stay in place. The sand and gravel are examples of soft coastal management, whereas the concrete blocks and the stones are hard coastal management. When there is a huge wave, for example during a storm, the pile of concrete works as a breakwater (see Section 1). When the wave hits the pile, it ''breaks'' and its strength and speed is decreased enormously (see picture). The rest of the wave will be stopped by the layers of boulder and sand. When the wave hits these layers it actually takes a ''bite'' out of the layers; part of the layers flush away, but because there is a big pile of concrete, the boulders and sand do not disappear into the ocean. This means that this coast will always be maintained. This is a very good example of ''dynamic coastal defense''.

The terrain height is 5,5 metres above sea level where the container and distribution activities are. In combination with the sea walls it has been calculated that this part of Maasvlakte 2 is well protected and complies with the standard of 1 in 10.000 years risk chance, even under very severe circumstances. The most suitable height for the locations where chemical products are stored has not yet been determined. It is possible that they will be placed at an even higher level because the risks will be much worse, when these locations are flooded.

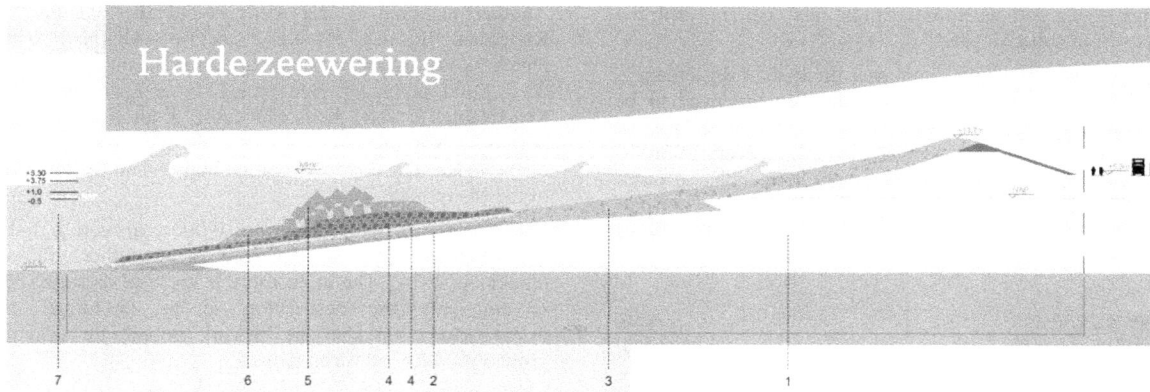

Figure 22: A drawing of what the coast and coastal defence at Maasvlakte II looks like.

4.6   The BOS-system

We already discussed some of the major storm surge barriers that defend the Port of Rotterdam and the surrounding areas, for example the Hartelkering and the Maeslantkering. It is very important that these storm surge barriers close immediately when a dangerous situation is about to occur to ensure that the barriers are closed in time. It may sound strange, but humans just cannot do this. The risks that humans do not see the danger coming or that humans close the storm surge barriers too late, are just too high. Therefore, the Dutch designed the BOS-system (Beslissings Ondersteunend Systeem = Decision Supporting System). This is a very advanced computer system that is applied at, for example, the Maeslantkering. The water around the Maeslantkering is filled with all kinds of sensors which detect the water level. When the sensors measure a water level value of 3 metres (= 9,8 feet) above average or more, the BOS-system, a computer system, ensures the Maeslantkering automatically closes. This system offers much more safety and this computer system virtually never makes any mistakes and is therefore much more reliable than humans. The BOS-systems consist of a lot of computers which are located at buildings on both sides of the Maeslantkering (Jonkman, 2014).

Figure 23: One of the two buildings with computers of the BOS-system.

As already mentioned, when water levels of +3 metres have been detected, the Maeslantkering closes (this happens approximately once every 10 years). This closing of the Maeslantkering takes place as follows: the docks at which the two parts of the Maeslantkering are rested when it is opened are filled with water. Because of this, the two ''doors'' are pushed into the water of the Nieuwe Waterweg (The New Waterway). The doors come really close to each other and then we call the Maeslantkering closed. When a (huge) storm wave approaches from the sea, it will be stopped by this barrier. There is a ''but'', the Maeslantkering is not actually entirely closed. This is because there is also water drainage from the Rhine towards the North Sea. If the Maeslantkering were to be 100% closed, all this water would accumulate. This causes a high water level and then the port and city of Rotterdam would still be flooded. However, when the Maeslantkering is not entirely closed, the Rhine water can still be drained towards the North Sea and at the same time it is made sure that waves from the North Sea will largely be stopped.

4.7   Maintenance of the coastal defence

The types of coastal defence used at the Port of Rotterdam are of very high quality, but they still need to be checked and when necessary repaired every now and then. The soft coast needs to be repaired more often because it is not as strong as the hard coast. The most important factor that affects the coastal defence is the waves and their power. Concerning soft coast, also the wind can be destructive because (dune) sand does not weigh much. Sometimes, a heavier layer of clay is placed upon the dunes that will not be blown away. The clay also makes sure that the water is stopped and does not flow into the dunes. When this happens the consequences can be disastrous. A wet dune is very unstable and can collapse. But, unfortunately, when the clay is applied, another problem arises. Now, the water will flow underneath the dune. The water will wash away sand from the bottom of the dune and now, the dune can also collapse.

This means that a situation in which no problems will appear can never be created. The dunes always need to be repaired and maintained by humans with, for example, beach nourishment.

A way to solve the last issue, the collapse of dunes, is placing heavy weights at the inland side of the dune. These

165

weights function as counterweight and can prevent the dunes from collapsing.

The hard coastal management near the Port of Rotterdam is of very high quality. Therefore, they do not need to be repaired very often. For example the upper stone layer of many dikes needs to be replaced only once every 30 or 40 years. Because the stone layers are not permeable, no water can flow into the dikes and so cannot affect the interior part of the dike. Because of this, the hard coast requires relatively little maintenance.

Figure 24: In December 2013, a dike near Schoorl, North Holland, collapsed.

4.8   Research

The total risk of an area can be shown in the diagram below.

Total risk = Chance of a flood x times in 10.000 years x impact in Euros (€).

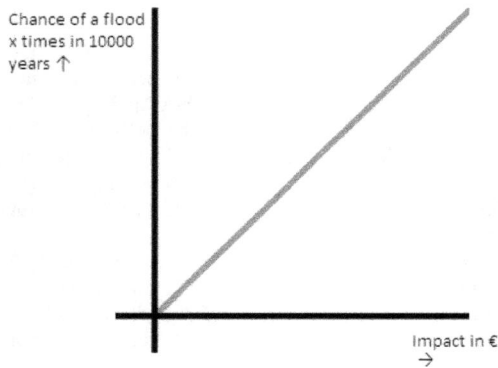

Figure 25: The chance of a flood and the associated impact in Euros.

The total risk is the cost of all the damage being caused by a flood multiplied by the chance of a flood in x times in 10.000 years (Jonkman, 2014). For example, the area around the Maasvlakte 2 has a chance of a flood of 1 in 10.000 years. Because the Maasvlakte 2 is not connected to the Netherlands, the only damaged that can be caused is to the Maasvlakte itself. The total risk will not be too high.

Another example is the area around the old Port of Rotterdam. Because this area is below NAP (NAP is the Dutch North sea level, as explained further on), the chance of a flood is bigger (see table). In comparison to the Maasvlakte, a flood here can cause a lot more damage because it is in the centre of one of the biggest cities of the Netherlands. The total risk will be higher than the total risk of the Maasvlakte 2.

The chance of a flood x times in 10.000 years can be 1, 10, 100 or more. This is explained in the other graphics (Jonkman, 2014). The impact in € is the total damage. This is not only the destruction of the buildings and infrastructure, but also the loss of ground, for example ground used for industrial crops.

NAP	Chance of a flood in one year
<0-0	1
+1	0,1
+2	0,01
+3	0,001
+4	0,0001
+5	0,00001

Figure 26: Chance of a flood in 1 year.

The chance of a flood can be determined in the following way. NAP means 'Normaal Amsterdams Peil', in English also known as Amsterdam ordnance datum. The NAP is almost the same as the level of the Dutch North Sea. When the NAP is 0, it means that the land is on the same level as the sea. If there were not any protection against the sea, there would be a constant flood. The higher the land is, the lower the chance of flooding. For example the dikes at Rozenburg are at a height of +1 NAP and therefore, the chance of a flood in one year is 0,1, which means one in ten. The dikes at Hoek van Holland have a height of +12 NAP which means that there is almost no change of a flood in a year.

Another calculation is also very important when deciding where to build a particular kind of coastal management. The calculation made is showed in the diagram. At the x-axis, you can see "height" in metres. This stands for the height of the dikes or storm surge barriers that are built. The higher the dikes, the more money it costs. But at the same time, when there is a flood, less water will get over the dike when it is higher. So when the dikes are higher, the damage during a flood will be lower. This is shown by the two black lines in the diagram the red line being the cumulative. At a certain height, there are certain building/maintenance costs and damage costs. When you count this number up at different heights, you can draw the cumulative. When deciding how high a dike, barrier etc. should be, researchers draw this diagram. They look at the point where the total costs are the lowest, look at the arrow, and decide the dike, barrier, dam etc. should be that high.

166

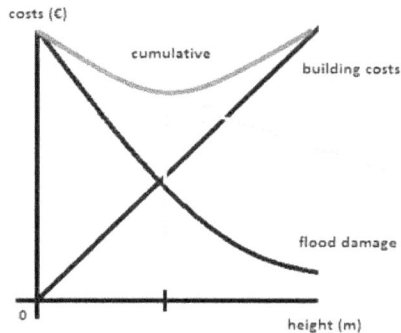

Figure 27: Costs and heights

4.9 Improvements about coastal management

The technical university of Delft offers several studies, for example coastal management, that focus on new interventions and changes within the coastal zone (Jonkman, 2014). They have an alliance with Deltares and together they work on multiple projects to improve the coastal defense. One intervention that is already helping the Maasvlakte is the 'fake' sea soil which breaks the waves as explained before. Because of the blocks they placed in the sea, the waves break before they hit the edges of the Maasvlakte, so there is not much weathering.

Another project they have been working on, is the sand motor, also called sand engine. Figure 28 shows the northwest coast of the Netherlands, including the area around Rotterdam. The sand shown in the figure has been brought there by the scientists. They have predicted that this sand will spread out over the entire coastline in several years. This means that the dunes will become bigger so they will protect the land better against the sea.

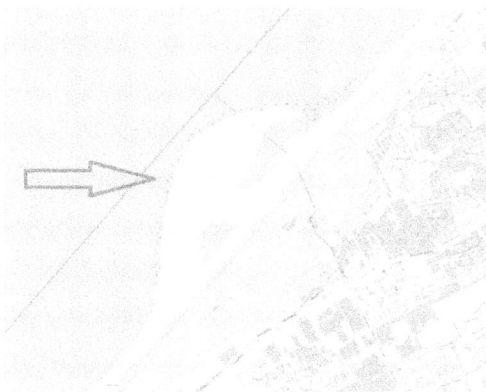

Figure 28: Zandmotor (Sand engine)

The Dutch are also known in the water business because of the IJkdijk. The IJkdijk is a facility which tests dikes. It is a big dike which is full of sensors. It is used to detect weaknesses in dikes. The concept of the IJkdijk is that in the future the results of the IJkdijk can be used to design warning systems for real dikes. Today there is only one IJkdijk, located in the Dutch province of Groningen.

Deltares offers a big centre where they can copy the (Dutch) sea. Because everything is artificial, they can accelerate or slow down processes. The photos show a part of the centre.

Figure 29: (Top) An overview of the Deltares research centre. (Bottom) A mini-dike has been built.

5 Future of the Port of Rotterdam

Now that we have a good impression of the existing techniques to defend the Netherlands against the water and we have taken a closer look at the Port of Rotterdam, it is time to look at the future of the port. At the moment it is defended properly, but the world is changing and therefore the Port of Rotterdam must also change. In this section, we will look at the weaker parts of the port. Where and what are its weaknesses? We will also take a closer look at new developed techniques to defend ourselves against the sea and their applicability to the port, the rising of the sea level and the advantages of a dynamic coastal defence.

5.1 Weaknesses of the port

The Port of Rotterdam is divided into two separated areas: one that is located within the dikes and one that is located outside the dikes. (Jonkman, 2014) The main part of the port is situated outside the dikes. When these parts of the port were built, the Dutch government already knew about the phenomenon of the rising of the sea level. Therefore, the most recent port areas are located high enough to keep up with the sea level rise in the next few years. The older parts of the port were built before anyone realized that the sea level was rising. These areas are not adapted to the greenhouse effect. Therefore, these are low-lying areas. And the closer an area is located to the sea level, the more vulnerable it is. Fortunately, not all of the old areas are still in use but the parts were there is industry, are obviously the weaker spots of the port.

Figure 30: One of the old parts of the Port of Rotterdam.

5.2   The sea level rise

The weaknesses of the Port of Rotterdam are just one of the concerns. Another, generally known problem, is the rising sea level. As the water level continues to rise, the port has to anticipate to that. On average, the sea level rises about 25 to 30 centimetres per century. (Jonkman, 2014) As the new parts of the port have already been modified to the increase of the water level and have a standard of being higher than two feet or more above sea level, these parts have a margin to deal with the rise. But nature is not reliable. It may happen that the sea level suddenly rises faster in the next few years. To be one step ahead of nature, we use measurements in the North Sea to predict the rise in sea level. We make models of the measurements so we can notice changes early. When the changes are detected at an early stage, we can take action in time. This is necessary because the increment of the port areas and the dikes will cost about 10 years. If the changes in the rise of the sea level are noticed too late, it could have detrimental consequences for the Port of Rotterdam.

5.3   New techniques

To keep up with the changes in the sea level and other circumstances, we need to develop new techniques to defend ourselves against the sea. This is not easy. Before a new technique is discovered, there is a large process of testing and measurement to make sure that the technique is good enough to use in the coastal defence. This takes a lot of time and money. (Jonkman, 2014)

The TU Delft, a technological university in the Netherlands, is one of the organizations that are involved in water management and coastal defence. They are constantly trying to find new ways to defend people against the water. This happens at different scales. The university itself has a laboratory where they can simulate waves in a large container by using large panels. In this way, the sea can be simulated. That is a good way to do several small experiments with regard to dunes and dikes. Furthermore, the TU Delft has invented the breakwaters, which are now used in the defence of the Maasvlakte.

The university also takes a close look at the weak spots in the dunes and dikes at the moment. They are doing research into the so-called failure mechanisms of the coastal defence. By performing simulations, the researchers can see where improvements are needed with regard to the dikes and dunes. All these developments are not only done by the university itself, they receive a lot of help from other companies like Deltares, a very important organization in the field of innovative developments related to water management.

5.4   Dynamic coastal defence

In the defending of the Port of Rotterdam, hard coastal defence is generally used. However, there are areas of the port that are structured in such a way, that they fit perfectly into the landscape. This is called 'Ecoshape'. (Nauta, 2014) People choose to build along with nature. This principle can for example be found at the Maasvlakte 2. The sides of the Maasvlakte which the water not immediately hits, where there is no breakwater, mostly consist of soft coastal defence mechanisms.

Figure 31: Maasvlakte II and dynamic coastal defense

Building with nature has some advantages. It is cheaper than the use of hard coastal management. It also adds a higher value to the appearance of the port, that is to say, it looks nicer and it is more attractive to live close to the port. Furthermore, any expensive maintenance is avoided by using 'Ecoshape' because the defence is built in harmony with nature. In fact, nature is doing a lot of the maintenance. It does all the important work. This saves a lot of money. Building a dynamic coast does not only benefit nature by integrating the coastal defence in nature but is it also good for the economy of the country.

5.5   The Port of Rotterdam in 200 years

It is difficult to say what the Port of Rotterdam will look like in about 200 years (Jonkman, 2014). There is no doubt that the port will continue to grow as it is an important trading post for the Netherlands and for Europe. Most goods are distributed in Europe by use of the Port of Rotterdam. Fortunately, there is enough space in the port to expand, so it can still grow. With the recent construction of Maasvlakte 2, people have created space for new businesses.

We do not think that a third Maasvlakte will be built any time soon. First, the recently built Maasvlakte 2 should be filled completely with companies and industry. The old parts of the port, that are unused at this moment, can also be filled up with new businesses. However, before that can be done, these parts will need to be adapted to make them safer. As mentioned before, the old parts of the port are situated at a very low level, around sea level, and because of this the flood risk is very high.

Another factor that makes it difficult to predict what the port will look in a few years is the rise in sea level. Because of the greenhouse effect it is possible that the rise in sea level will increase sharply in the coming years. To be prepared for that, the port will have to take important measures. But if the rise continues at the current pace, with an increase of 25 to 30 centimetres a century, major measures are not necessarily needed. The TU Delft is one of the companies that are concerned with the prediction of the future of the Port of Rotterdam. With models and simulation programs researchers try to make vistas and visions of the future about the port. If the sea level rises increasingly, there will be a good chance that the port will become a kind of polder. The dikes will be raised but the terrain behind those dikes remains at a lower level. In this way the defence of the port is not too expensive but there will still be a good protection. It is to be expected that more industry will be located outside the dikes in the future.

Figure 32: In the future, Maasvlakte II will be totally covered with industry

The future of the Port of Rotterdam is recorded in the 'Havenplan 2020' (Port Plan 2020). This plan was formed by the local authority of Rotterdam in September 2004. The main objectives of this plan are to strengthen the international position of the Port of Rotterdam; to increase the economic structure in Rotterdam and to increase the spatial quality of Rotterdam and the port to make it more liveable. The plan includes ideas to make the port sustainable. Building along with nature is one of these ideas. They want to create an ecological area on the south side of Maasvlakte 2, where nature is the most important thing. As mentioned earlier, building with nature does not only have positive effects on the appearance of the port but it is also better for the economy as it is cheaper. (Havenbedrijf Rotterdam N.V., 2004)

In short, the weakest part of the port is currently the old port area which is still used by companies. The weakness is due to the low position with respect to the water surface. One of the suitable solutions for this part of the port is to make a polder of it. The dikes around the old port area increase but the port itself does not increase. It remains at its low location. By using this method, a lot of money can be saved. Another option is integrating the port in nature by increasing the use of the principle of 'Ecoshape'. The result of this would be that the port looks nicer and it would also be cost-effective.

6 Water management in Singapore

The port of Singapore is, after two other ports in China, the biggest port in the world. The port is an absolute necessity for Singapore since Singapore's surface is not very big and the country itself also does not possess many natural resources. The port makes it possible for Singapore to import natural sources, of which a large part will be processed and later exported. The port has experienced a lot of growth in recent years. This, of course, also brings along some challenging situations. We thought it would be interesting to compare water management (in the port of) Singapore to water management in the Netherlands.

Figure 33: Singapore harbour, a very extensive and complex system

6.1 The challenges Singapore faces

The number of inhabitants and the economy in Singapore are growing rapidly, but Singapore only has a surface of 638 km2 (= 396 mi2) and an estimated 3000 ha (Nauta, 2014) will be needed, so actions have to be taken. Singapore already expanded itself a great deal by land reclamation. Changi, Tekong, Jurong Island and Tuas area are examples of Singapore's reclaimed land. Figure 34 is slightly outdated (2003), but it shows the growth of Singapore after the 50s of the preceding century very well. The reclamation of land increased Singapore's land area by 17% (Wild Singapore, 2013).

Singapore now reclaims land, simply put, by dropping large amounts of sand on the place where the land is determined to be. This way, the reclaimed land lies above sea-level, which of course is favourable. This could be compared to the way "Maasvlakte 1" and "Maasvlakte 2" have been established in the Netherlands. However, sand is not endlessly available. Singapore used to import sand from Indonesia, but Indonesia could no longer cope with the demand and raised the price severely. That is why other options will have to be explored. Polders are one of them. Polders are created by the reclamation of land embanked by dikes and widely used in the Netherlands. Polders do often lie beneath or at sea level, but are relatively safe when the surrounding dikes are constructed well. In addition, this way the amount of sand that has to be used will be minimalized.

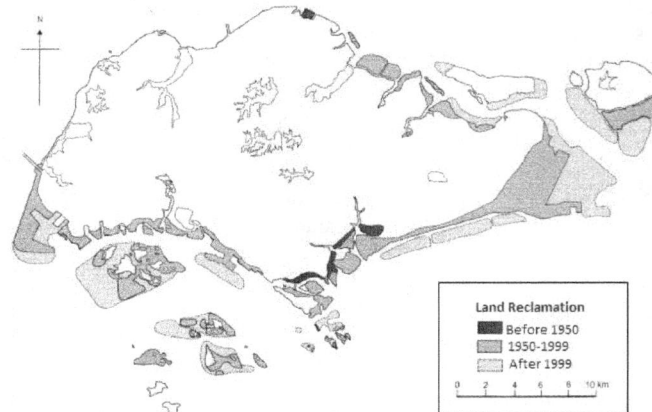

Figure 34: Singapore's land reclamation up till around 2003

Singapore will probably have trouble producing enough freshwater (drinking water) in the future (Nauta, 2014). Right now, there are three main sources used for the production of drinking water: Sea water (reverse osmosis), water that has already been used (water recycling), producing fresh water by the closing estuaries and import from Malaysia. It has been found these sources may not suffice in the future and Singapore also wants to be able to function independently from Malaysia. This is why Singapore closes estuaries in order to produce fresh water, but also in order to protect their coast against the sea. Marina Bay is an example of such an estuary, the dam is known as the Marina barrage and was finished in 2008. The Marina barrage is more or less comparable to the closures of the estuaries (part of the Delta project) in the Dutch province Zeeland after the flood of 1953. It is a fast and practical solution, but it also has a downside, the ecosystems in the estuary and in the land surrounding the estuary are used to salt water flowing in from the sea, thus a huge difference. A lot of effort will have to be put in in order to restore the ecosystems that will be damaged. However, safety and water production are also important objectives so it is a delicate consideration.

The closure of an estuary is of course not the only radical change made to the coastal area of Singapore. The land reclamation along the coast is also part of this. These changes will have big effects on the flora and fauna living in the (surrounding) areas. For instance, some animals that were at the top of the food chain have disappeared causing imbalance in the remaining habitat, which in turn can lead to the disappearance of other species and eventually create a significantly less diverse flora and fauna in the concerning areas. For example, tigers used to live in the mangroves, now there are not any tigers anymore. Animals such as crocodiles are also no longer common and dolphins, dugongs and sea turtles are also less commonly seen.

6.2 Incorporating nature in the port of Singapore

Building with nature is not (yet) a common practice in Singapore, whereas in the Netherlands there is already a lot of consideration put into the environment. Although it is of course not always possible or profitable to make big changes, especially in the area of the port some more nature

could be incorporated, since even small changes can make a big difference.

There are a lot of ways to improve the Eco friendliness of water defences. One of which is using what the company Deltares has named 'the Pile hula'. Pile hulas are very useful, especially in ports, because concrete columns can often be found here. Pile hulas are nylon strips that look like Hawaiian skirts when they are attached to columns or piles. Within a relatively short amount time after the attachment they will be habited by all sorts of little life forms such as mussels, barnacles and a range of algae. According to a pilot study in the Port of Rotterdam, an average of 8.5 times more biomass is found on the pile hulas than on ordinary piles (Deltares, 2014). The shellfish are also important because they are able to filter the water by collecting food for themselves.

Another way to build with nature would be the pontoon hula, this is a variation on the pile hula. They work in almost exactly the same way, but instead of attaching it to a pole, the pontoon hula is a floating structure made from PVC. Other than giving home to small creatures, Deltares has found these pontoon hulas are also able to damp reflection waves in harbours.

A third solution could be changing the form of the coating of hard coastal defences and thereby making sure water can be retained in higher parts of intertidal areas. This is a cheap change which can result in an enormous boost for local biodiversity. Eco concrete is more or less the same idea, it is concrete with a rougher surface, a special texture. It allows organisms such as algae, seaweed, periwinkles and mussels to colonise it more easily. Mussel colonies are able to filter the water.

In Rotterdam, soft coastal management is incorporated in the Maasvlaktes. The Western side consists of mainly hard coastal defences, but the eastern side also has soft coastal management incorporated, making the water defences more eco-friendly than when using only hard coastal management. As said earlier, the Maasvlakte can be very well compared with Jurong island or eventual other comparable islands. It could be possible to incorporate both hard and soft coastal management here. On the side where the wind and the water smashes against the shore, hard defences could be used, on the other side it might very well be possible to use soft defences.

Mangroves are also a solution, not in the port of course, but more in the direction of the North, as Singapore is owner of some very nice mangrove forests. These forests are natural coastal defences and also very valuable ecosystems. Often, hard coastal defence is used in order to reduce coastal erosion and improve the coastal protection. However, on the terrain where these mangroves are located, hard defences are often not sustainable because they subside quickly. Instead of getting rid of the mangroves, other solutions can be found such as building berm structures from natural materials in front of the coast.

Conclusion

After doing a lot of research and getting in contact with the right people, we were able to complete the six sections. We wrote these sections to eventually answer our main question:

"What is the best way to defend the Port of Rotterdam against the sea?"

In the first section, we stated that it has always been very important for Rotterdam and the surrounding areas that it is defended well against the sea, since almost all of this area lies below sea-level. Numerous times, (the area of) Rotterdam flooded and many lives were lost over and over. After the terrible flood of 1953, the Dutch government decided that a proper coastal defence should be created; the Delta plan was made and brought forth the Delta works. These storm surge barriers also defend the city of Rotterdam. But we found that the Port of Rotterdam is not mainly defended by the Delta Works or other dikes. In the third section it is stated that this is so because if the Port of Rotterdam would be situated behind the dikes, other measures would have to be taken in order to have the port operate well.

As you could read in section two, about hard and soft coastal management and in section four, about the defence of the harbour, mostly hard coastal management is applied at the Port of Rotterdam; like for example the Maeslantkering. We would highly recommend that other means of coastal defence are considered to apply at the port. This is because we found out that hard coastal management can have very negative consequences for components such as nature and environment. Hard coastal management is not even closely as dynamic as the sea and the shore and is therefore, according to us, not the absolute best way to defend the Port of Rotterdam. We recommend that more dynamic defence mechanisms, such as beach nourishment, should be used to defend the Port of Rotterdam. We know this will cost more money, but we find that soft coastal management will secure a healthy environment. We also think that soft coastal management offers more advantages for the future. We don't know what the future brings; maybe the sea level rises a lot or the soil at the Port of Rotterdam sags. It will probably be easier to adjust soft coastal management techniques to these kind of changes. We simply cannot adjust huge buildings such as the Maeslantkering to certain new conditions.

We are happy to say that the new areas of the port, for example the Maasvlakte, are defended in a relatively dynamic way. As described in section four, the Maasvlakte is defended by a mixture between hard and soft coastal management. Therefore, the coast of the Maasvlakte is one of the most dynamic parts of the Port of Rotterdam and we are glad to see this development. Still, there are things we are worried about. The old parts of the Rotterdam harbour are still vulnerable. This was stated in section 5, ''Future of the Port of Rotterdam''. These parts are at low level and therefore, have a significantly higher flood risk than other parts of the port. Even though not all the old parts of the port are still in use, we would like to make some recommendations about how to make these parts safer. One way to ensure this, is to make a polder of the old parts. This means to create (higher) dikes around the areas. The parts themselves stay at the same height and this saves a lot of money. We encourage the use of natural materials as stones, sand and clay to make the dikes. We discourage the use of static, non-dynamic, hard materials such as concrete.

A last recommendation we want to make, has to do with Singapore. We are amazed to see how fast the Port of Singapore has grown and it still is. However, we are concerned to see that a lot of estuaries have been closed to obtain drinking water and that mostly hard coastal management is used. The Dutch also closed their estuaries when building the Delta Works and this caused major fish mortality. Other than building more soft coastal defences (which we know is difficult to apply in Singapore, as there is a sand deficit) other, smaller changes could be made in order to preserve nature such as the use of pile hulas at the coast. According to studies done by Deltares, these concrete columns attract shellfish and they can filter the water. Using pile hulas, nature can somewhat be incorporated in the port of Singapore. A second advice concerns the hard coasts used in Singapore. We think the knowledge from the Port of Rotterdam can be used in a great way in Singapore. There they could also incorporate the dynamic coasts as used in the Maasvlakte. By these means, the Port of Singapore will be safe and secured and with the least consequences for the natural environment.

Acknowledgements

In this report, we inform the reader about everything that has to do with the Port of Rotterdam: about its history, about how it is defended against the ruthless sea. A very important point in this report was discussing what could change in the future. Which parts of the port should be protected better etc.? In section 6, we mentioned the huge Port of Singapore. This is a growing port and therefore, one always has to be on the lookout for problems. To secure a prosperous future, we recommend that research about the impact of the growth on both the economy and ecology be done. Furthermore, we are glad to say that the majority of the Port of Rotterdam and surrounding areas are defended properly which makes it is a safe place to be. Only the old parts of the port could be defended a little better, as they are situated at a relatively low level. Therefore, the flood chances here are higher than elsewhere in the port. We recommend that a research on the old parts be done so that the safety of these parts will also be secured in the future; a future with a rising sea level.

We would never be able to conclude these things without the professional help of Prof. Bas Jonkman and Tjitte Nauta. Prof. Jonkman, professor of Integral Hydraulic Engineering at the Delft University of Technology, told us everything we needed to know about the Port of Rotterdam. We are also very grateful that Mr. Nauta could tell us a lot about the Port of Singapore. Together, they helped us to complete an extensive study.

References

Internet

[1] Aantjes, H. (sd). Levees, dikes and water defences. Retrieved in january 2014, from Deltares: http://www.deltares.nl/en/expertise/100801/levees-dikes-and-water-defences

[2] Baptist, M., & Wiersinga, W. (2012). Zand erover: vier scenario's voor zachte kustverdediging. Retrieved in january 2014, from Wageningen UR: http://www.wageningenur.nl/nl/Publicatiedetails.htm?publicationId=publication-way-343233323536

[3] Beerda, E. (2012, april 5). Zachte kustverdediging voordeliger. Retrieved in january 2014, from CoBouw: http://www.cobouw.nl/nieuws/algemeen/2012/04/05/zachte-kustverdediging-voordeliger

[4] Deltares. (2014, january). Magazine Delta Life. Retrieved in march 2014, from Deltares: http://www.deltares.nl/nl/over-deltares/magazine-delta-life

[5] Dredging Today. (2013, july4). VIDEO: Land Reclamation in Singapore from 1984 to 2012. Retrieved in march 2014, from Dredging Today: http://www.dredgingtoday.com/2013/07/04/video-land-reclamation-in-singapore-from-1984-to-2012/

[6] Havenbedrijf Rotterdam N.V. (2004, september 16). Havenplan 2020. Retrieved in march 2014, from Maasvlakte 2: https://www.maasvlakte2.com/kennisbank/200409%20Havenplan%202020.pdf

[7] Havenbedrijf Rotterdam N.V. (2014). Factsheet harde zeewering. Retrieved in march 2014, from Maasvlakte 2: https://www.maasvlakte2.com/uploads/factsheet_harde_zeewering.pdf

[8] Rijksoverheid. (2011, juni 7). Bestuursakkoord Water. Retrieved in january 2014, from Rijksoverheid: http://www.rijksoverheid.nl/onderwerpen/water-en-veiligheid/documenten-en-publicaties/rapporten/2011/06/07/bestuursakkoord-water.html

[9] Rijkswaterstaat. (2010). Jaarverslag Rijkswaterstaat 2010. Retrieved in january 2014, from Rijkswaterstaat: http://www.rijkswaterstaat.nl/figures/Rijkswaterstaat%20Jaarverslag%2020

[10] Rijkswaterstaat. (sd). Hartelkering. Retrieved in march 2014, from Rijkswaterstaat: http://www.rijkswaterstaat.nl/water/feiten_en_cijfers/dijken_en_keringen/europoortkering/hartelkering/

[11] Rijkswaterstaat. (sd). Maeslantkering. Retrieved in march 2014, from Rijkswaterstaat: http://www.rijkswaterstaat.nl/water/feiten_en_cijfers/dijken_en_keringen/europoortkering /maeslantkering/

[12] Rooijen, D. v. (2005). The northern sea defence of Maasvlakte 2. Retrieved in february 2014, from TU Delft: http://repository.tudelft.nl/view/ir/uuid:caa39965-6dad-4194-aef3-62699c7c1c40/

[13] RTV Noord-Holland. (2013, december 6). Blijdschap om duindoorbraak in Schoorl. Retrieved in march 2014, from RTV NH: http://www.rtvnh.nl/nieuws/128301/Blijdschap+om+duindoorbraak+in+Schoorl+(nu+met+video)

[14] Tan, R. (2013, august 8). Lost of coastal ecosystems. Retrieved in march 2014, from Wild Singapore: http://www.wildsingapore.com/wildfacts/concepts/loss.htm

[15] Vliz. (2004-2014). Onze kust... Retrieved in january 2014, from Vliz: http://www.vliz.be/vmdcdata/faq/question.php?qid=125

[16] Walle, B. V. (sd). Kustverdediging: de strijd tegen de zee. Retrieved in january 2014, from Vliz: http://www.vliz.be/docs/Groterede/GR03_strijd.pdf

[17] Wengel, T. (1997). Hard, zacht, of een combinatie? Retrieved in january 2014, from TU Delft: http://repository.tudelft.nl/assets/uuid:0daac9df-5724-453b-b46e-f88cebe5e81d/terWengel1997.pdf

[18] Wikimedia Foundation. (2014). Beach nourishment. Retrieved in january 2014, from Wikipedia: http://en.wikipedia.org/wiki/Beach_nourishment

[19] Wikimedia Foundation. (2014). Dam (waterkering). Retrieved in january 2014, from Wikipedia: http://nl.wikipedia.org/wiki/Dam_(waterkering)

[20] Wikimedia Foundation. (2014). Deltawerken. Retrieved in january 2014, from Wikipedia: http://nl.wikipedia.org/wiki/Deltawerken

[21] Wikimedia Foundation. (2014). Dijk (waterkering). Retrieved in january 2014, from Wikipedia: http://nl.wikipedia.org/wiki/Dijk_(waterkering)

[22] Wikimedia Foundation. (2014). Flood control in the Netherlands. Retrieved in january 2014, from Wikipedia: http://en.wikipedia.org/wiki/Flood_control_in_the_Netherlands

[23] Wikimedia Foundation. (2014). Polder. Retrieved in january 2014, from Wikipedia: http://nl.wikipedia.org/wiki/Polder

Figures

Wikimedia Commons. User: Vladimir Siman (Oosterscheldekering), Weitbrecht (naval ship worm), Mohammed F.M. Yossef (groynes), M.P. Tillema (the dike), Rijksmuseum Amsterdam (painting), Rijksdienst voor Cultureel Erfgoed (Kinderdijk). Rijkswaterstaat (Maeslantkering), deltawerken.com (map Hartelkering), engineersonline.nl (Hartelkering), refdag.nl (queen Beatrix), Google Maps (satellite Maeslantkering), RTV NH (collapsed dune), AsiaNews (Singapore harbour), Wild Singapore (land reclamation), Havenbedrijf Rotterdam N.V. (map Maasvlakte II).

Appendix

Explanatory word list

-	*Brug (used in Zeelandbrug):*	Bridge
-	*Meer:*	Lake
-	*Noord:*	North
-	*Oost:*	East
-	*Rijkswaterstaat:*	Ministry of infrastructure and environment, the ministry
		Takes care of water management in the Netherlands amongst others.
-	*Schelde (used in Westerschelde):*	Scheldt; a river, the Western Scheldt is the estuary of the Scheldt.
-	*Sluis (used i.e. in Bathse Spuisluis):*	Sluice
-	*Tunnel (used in westescheldetunnel):*	Tunnel
-	*Vlakte (used in Maasvlakte):*	Plain
-	*(Water-)kering:*	(Water-) defence
-	*West:*	West
-	*Weg:*	Way, often used in Waterway (Dutch: Waterweg)
-	*Zuid:*	South

Geographical places

-	*Botlek:*	Harbour and industrial area in Rotterdam.
-	*Groningen:*	A Dutch province, situated in the North of the Netherlands.
-	*Hoek van Holland:*	Literally 'corner of the Netherlands'. It is a town located next to the sea in the province Zuid-Holland.
-	*IJsselmonde:*	Isle in Zuid-Holland.
-	*Kinderdijk:*	Village in Zuid-Holland ca. 15 kilometres east of Rotterdam.
-	*Randstad:*	Most economically important and densely populated part of the Netherlands. It covers a part of Noord-Holland, Zuid-Holland and Utrecht. Rotterdam is also located here.
-	*Zeeland:*	Province of the Netherlands located in the South-West.
-	*Zuiderzee:*	Former part of the North Sea that is now separated from it by the 'Afsluitdijk' or, in English, enclosure dam. After the closure it was given the name IJsselmeer (lake IJssel). It is the largest lake in the West of Europe.
-	*Zuid-Holland:*	South-Holland. Province of the Netherlands located in the West.

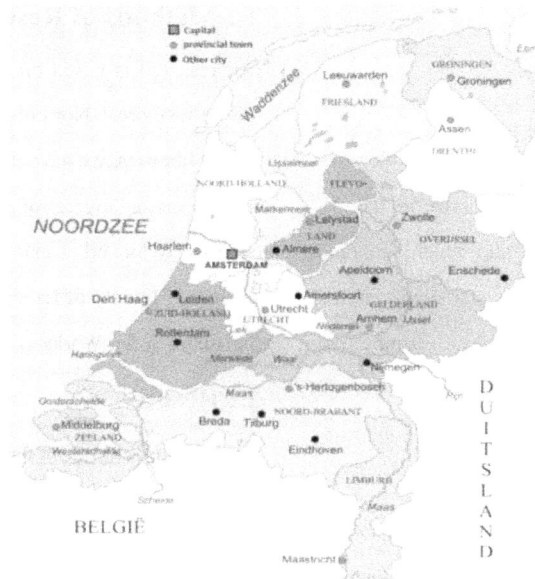

Figure 35: Map of the Netherlands with the Capital and other big cities indicated

Wasting Water in Japan: How Countries Can Conserve Water

Wakako Narukama, Karin Yoshida, Rick Saito

Shibuya Senior High School, Japan, natsume@shibuya-shibuya-jh.ed.jp

Abstract

Since the end of WWII, Japan has been relying on imports of oil, rare earth metals and other daily necessities. With food production rates declining and the labor force shrinking due to low birth rates and aging population, Japan will probably have to continue to depend on its imports. However, there is actually one natural resource that Japan provides on its own: water.

Our aim is to determine the factors that enable Japan to sustain its water resources, whether these factors are historical, geographical, political, and/or technological. Then, we aim to ascertain why Japanese people waste gargantuan amounts of water every day. Research indicates that the amount of water the whole population misuses in a day could supply the whole continent of Africa for one day. And to end on a positive note, we will analyze this data and devise a plan to reduce the amount of water squandered by Japan, a plan we hope could be a model for other developing and developed nations.

Key Words

virtual water, food, toilets, recycling

1.   Introduction

Water is deemed the most valuable resource in the 21st century. No matter by which country and which economic sector, water is consumed everyday. As water supply around the world decreases, countries have started to dispute over control of reservoirs and lakes as well as territories with underground water while corporations have been purchasing land to obtain its water supply.

Living in Japan, a stable, rich nation, not once have we heard in the news a story highlighting any issues about clean water. So, generally our team perceived that Japan had no issues with water and that we in fact wasted a lot of water. But, through this research, our assumption was proved completely false, in two ways: Japan actually does not waste as much water, but in fact does have a major issue concerning water.

2.   Content

2.1 Purpose of Research

Our team sought to ascertain whether Japan, while lacking in raw materials and dependent on foreign imports, is a country with an ample supply of water. In newspapers and on the news, we hear reports on Japan's dependence on imports of food and rare metals, but never water. So, our team conducted research to determine Japan's current water supply, the amount of water each sector uses, its geographical features such as rivers and forests that contribute to Japan's water supply, and its use of water as international aid.

2.2 Methods of Research

Our research is based on books, news articles, reports and school textbooks. Not only have we used local libraries and school libraries, we have used the United Nations University Library as a source of UN documents. We have also extensively looked at many websites, including those of the Japanese Ministry of Land, Infrastructure, and Transport and the Tokyo Water Works.

2.3 Results

Surprising findings on Japan's water supply

So it seems Japan is abundant in water supply and its people are not exactly wasting surfeit amount of water. Japan is an archipelago with four main islands and more than 3,000 small islands, covering a combined area of approximately

377,000 km[2]. The country lies in the northeast tip of the Asian Monsoon Zone, and the weather is generally mild with considerable variation from north to south. The country's four distinct seasons feature three periods of heavy precipitation.[1] This is during the tsuyu and typhoon seasons on the Pacific Ocean side, and during the typhoon season and in the winter on the Japan Sea side. Thus, it is often said that Japan's water resources are plentiful mainly during the tsuyu, typhoon, and spring thaw seasons.

Japan is also enthusiastic when it comes to aiding other countries with its water supplies. At the Fourth World Water Forum in 2006, Japan announced the Water and Sanitation Broad Partnership Initiative. (WASABI) Indeed, Japan's disbursements of aid for water and sanitation are the largest in the world.[2] The country provides expertise as well as equipment and facilities to other nations, based on its experiences, knowledge, and technology. One example is the Water Supply and Hygiene Improvement Project in Host Communities of Dadaab Refugee Camps. Located in Kenya's North Eastern Province near the border with Somalia, the Dadaab Refugee Camp currently holds 450,000 Somali refugees. With the constant increase of refugees, the camp witnessed a grave deterioration of environment and public order. To solve this disparity, Japan decided to assist the Kenyan host community with their water supply, and has constructed deep well water supply facilities and reservoirs, supplied water trucks, and provided training aimed at improving the maintenance, management, and sanitation of water supply facilities. Japan hopes that its efforts to improve hygiene and supply enough water for the host community's residents will help to solve the problems within the camps.

However, through are research, we have discovered that based on the theory of "Virtual water," Japan is in fact low in supply of water to meet the people's demands.

What is "Virtual water?" Introduced in 2008 by Professor John Anthony Allan from King's College London and the School of Oriental and African Studies, this concept is that when goods and services are traded, there is also a hidden flow of water. [3]Although there is no direct tradeoff of water, water is used to produce these goods and therefore is taken out of the countries producing these commodities. It follows that the country importing these goods would not have to use any water to obtain these goods.

Research indicates that to make one kilogram of wheat, 1000 liters of water are used. For one kilogram of corn, 1,800 liters would be needed. And for beef, it would take around twenty thousand times the amount of water.[4] For the country importing goods, this is a massive saving of its water supply.

"Virtual water" is detrimental to the county producing goods. Massive amounts of water that could be used for its own purposes are virtually shipped out of the nation. For developing nations, this could mean an increase in lack of access to clean water resulting in malnutrition and disease.

According to the Japanese Ministry of Environment, in 2005 Japan imported around 80 billion liters.[5] This is around the same amount the Japanese uses in a year. So, if Japan had to produce all of its food it imports, Japan would have to spend the same amount it already uses to produce food.

Why does this occur? The main factor that forces Japan to rely on foreign imports is Japan's low self-sufficiency rate. At just 40 percent, Japan cannot sustain its population with its current agriculture. As a result, Japan continues to rely on foreign products.

But the problem does not end here. After further research on Japan's food, we have discovered that loads of the imported goods are thrown away, not used at all. According to the

[1]http://www.japan-guide.com/e/e2277.html

[2] *Ministy of Foreign Affairs (2013)Japan's International Cooperation*

[3] Akio, Shibata, 柴田明夫 (2011). *Japan is the World's Number One Water Resource and Water Technology Nation*

[4] Akio, Shibata, 柴田明夫 (2012). *Japan: Taking Control with Water*

[5] http://www.env.go.jp/water/virtual_water/

Japanese Ministry of Food and Agriculture's 2007 study, 11 million 470 thousand tons of food in total, worth 11 trillion yen (110 billion dollars) was discarded[6].

In short, Japan would actually be low in water supply according to the theory of "Virtual Water" yet it wastes a gargantuan amount of food produced by this "Virtual Water" that was taken from exporting nations.

On Conserving Water

We would also like to show our findings on the topic of water conservation. Luckily, Japan has many ways of conserving water.

One significant example is the 'washlet', a nifty toilet created by TOTO with a heated seat and water spray features. Other functions include a bidet, blow dryer, seat heating, automatic flushing, etc. While the toilets in the 1970s used 20 liters per flush, the 'washlet', released in 1980, only used 8 liters. The earlier toilets used large amounts of water to clean, wash and send out, and close in the smell, of the toilet. TOTO succeeded in making these changes by using a smaller amount of water for the entire flushing process. Compared to the older version, this was equivalent to the amount used just to wash out the toilet. Since Japan did not have strict regulations on conserving water unlike countries such as the United States, TOTO realized that they could improve their products to be more ecologically friendly.

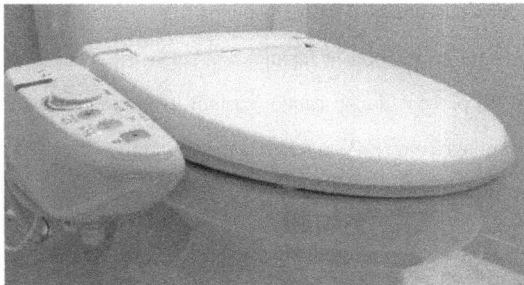

Figure 1: A typical Japanese "Washlet"

Now conserving 4.8 liters per flush[7], the unique product has been so highly praised that TOTO has begun selling it abroad in places such as California.

Another example would be a contemporary architect named Hikari Kurihara who has also contributed to push conservation efforts through his architecture. Influenced by the system for conserving water as major component of the Edo (the former name of Tokyo) home, Kurihara was inspired to create something similar. Rainwater, stored in a two-ton tank, is used for bathing, the flushing of two toilets, and the watering of a "green curtain" – a wall of vines to keep the scorching summer heat out of the house.

The famous Tokyo Dome has a rainwater collection and recycling system as well. Rainwater that falls on the 'membrane structure roof' is collected and stored in underground storage tanks.[8] While this helps prevent the overflow of water in sewers, the rest is used at the Tokyo Dome to flush toilets and store for fire prevention systems.

3.　Conclusion

Not only will Japan have to conserve its water but will Japan have to solve the "Virtual Water" situation.

More measures must be taken to increase Japan's self-sufficiency rate and decrease its dependence on foreign goods. If Japan continues to waste its imports and thus "Virtual Water" from other developing nations that could be using the water for their own purposes, criticism from the international community will soon follow.

As a way to substantiate the high demand for food, perhaps campaigns to raise awareness about "Virtual water" must be initiated to sway the public from excessive consumption. As Japan has a history of successful awareness campaigns, a plan to call into issue "Virtual Water" and Japan's overdependence on foreign water will have fruitful results.

There are multifarious solutions to this issue, whether they

6　Tomoko, Sakuma, 佐久間智子 (2008). *Water Business*

7　The World's Best Toilet: The Story Behind Creating the Washlet
8　http://en.g-forse.com/?eid=10

are technological or political. Japan and its corporations must continue to bolster their efforts to conserve more water for future generations.

4. Acknowledgements

Our team would like to show our most appreciation to Ms. Natsume, our guidance teacher who throughout this project supported our team. We would like to thank her for her outmost assistance.

We would also like to express our thanks to the libraries that we used for research. The libraries include: Shibuya Senior High School Library, Toshima-ku Central Library, and the United Nations University Library.

5. References

[1.] Black, Maggie and King, Jannet (2013)

The Atlas of Water （published by Maruzen）

"Tokyo Dome's Rainwater Collection and Recycling System | G-ForSE | Global Forum for Sports and the Environment." *G-ForSE*. N.p., n.d. Web. http://en.g-forse.com/?eid=10

[2.] Japan Guide (2006). Japan's Rainy Season (Tsuyu or Baiyu). http://www.japan-guide.com/e/e2277.html

[3.] Ministy of Foreign Affairs (2013)

Japan's International Cooperation

[4.] Ministry of Environment *Virtual Water* http://www.env.go.jp/water/virtual_water/

[5.] Ryosuke, Hayashi *林良祐* (2011). *The World's Best Toilet: The Story Behind Creating the Washlet* 『世界一のトイレ ウォシュレット開発物語』

[6.] Tomoko, Sakuma, *佐久間智子* (2008).

Water Business『ウォーター・ビジネス』 作品社.

[7.] Akio, Shibata, *柴田明夫* (2012).

Japan: Taking Control with Water『水で世界を制する日本』 講談社.

[8.] Akio, Shibata, *柴田明夫* (2011).

Japan is the World's Number One Water Resource and Water Technology Nation 『日本は世界一の水資源・水技術大国』 講談社

[9.] Kazunari, Yoshimura *吉村和就* (2012).

Book on Latest Trends in Water Business and its Intricacies『最新水ビジネスの動向とカラクリがよ〜くわかる本』秀和システム

Water Management In Austria

Amadia Kilic, Felix Oblin

Sir-Karl-Popper-Schule, Austria, martin.windischhofer@gmail.com

Abstract

A highly significant aspect to consider when investigating the importance of water in our daily life is the sustainable and responsible approach to water usage. Being the representatives of the Austrian state, we are highly aware of the fact that the Austrian state territory is advantageously situated in terms of water management and water preparation. Hence one of our projects' aims is to investigate the Austrian water management. Thereto we can enumerate topics as Austrian river basins, national techniques of water processing or the question of water privatization and marketing.

Accordingly the project should also compare and contrast the Austrian to the global situation in terms of water economics, resources or average water usage. We target to show how we could realize projects in an international community in order to enlarge our ecological awareness and help others with fewer resources to construct their own supply systems by sharing our know-how. Furthermore, since this project is presented to students, we also hope to be able to examine possible ways of saving water using our school facilities as an example and preserve a positive spirit.

Keywords

Education, Awareness, Water Management

1   Introduction

The project "Water Is Life" already began in 2013, when our school, namely the Sir-Karl-Popper-Schule, announced a competition. Students had to do research on the conference topic and in the following present the outcome of their investigation. The two winning projects both treat the topic "Water management in Austria". In order to give a complete image of our researches, we will combine them by creating a logical chain of thought – starting with examining the legal basis in Austria – continuing with the perception of water as a unique good – and ending with the creation of awareness for a sustainable water usage.

2   Content

2.1   The purpose of the investigation

We decided to inform ourselves at first about the laws settled. A legal basis is always a basis, which should be well known, before announcing revolutionary ideas or proposing futuristic plans. Only then, when encountering the amount of legal settlements, we realized how important the legal issue of water is. This new insight increased our awareness. We now comprehend that the first thing to start with when aiming for a sustainable water usage worldwide is exactly this key word "awareness". It is the key factor for every successive action, regardless of its type – research & science, daily water usage, water management, water economics and the like, it all depends on awareness. We delved into the science of how to raise people's awareness for water and, furthermore, how sustainable water usage can actually be put into effect.

2.2   Method of the investigation

To gain valid knowledge about raising awareness and sustainable water usage, we took advantage of the following research methods:

- Scientific books
- Researching websites and published documents, interviews and polls of
  - Water awareness projects & initiatives
  - EU (Water Framework Directive)
  - The Austrian Ministry of Life
  - Water management firms
  - Water facilities
  - Sustainability guides for individuals

Our research was thoroughly reproductive.

2.3   Outcome

It is the duty of every country's government to supply its inhabitants with access to potable water and sanitation. In order to fulfill this task a national organization is needed. Nowadays in Austria the state targets to collaborate with local inhabitants more intensively. Therefore, the "Flussdialog" project was established in 2008/2009 by the Austrian Ministry of Life. Since then 10 "Flussdialoge" have been carried out. It aims to create a dialogue between different interest groups. Mainly it gives the regional population the direct possibility to participate at water-related decisions, which will have an effect on their life.

Having a look at a map of the Austrian riversheds, we can easily see that international collaboration is needed.

Figure 1: Watersheds in Austria

Austria's state territory belongs to three international riversheds.

- "Danube-rivershed"- This is the most important rivershed in Austria as it covers 96,1% of the state territory. The Danube rivershed is unique with regard to various aspects. It is our world's most international river basin, meaning it is split up among 19 countries. In order to coordinate the interests of all these countries regarding water management, at the request of Austria, the IKSD (International Commission for Protecting the Danube) was established in 1999. Since 2000 it is also a platform to coordinate the implementation of the EU Water Framework Directive.

- "Rhine-rivershed" – This rivershed covers 2,8% of the state territory. The river is split up among 9 countries and the organization which deals with the coordination is called "IKSR (International Commission for Protecting the Rhine)"

- "Elbe-rivershed" – This rivershed covers only 1,1% of the state territory. The river itself is not in Austria. The commission IKSE (International Commission for Protecting the Elbe) coordinates the interests of 4 countries, whilst Austria only has only an observer status [1-4].

Before starting to talk about any water-related topic, the first and definitely most important awareness is the simple three words "Water Is Life" – also the conference motto. Water is essential and indispensable for every human being and therefore requires highest attention.

Living in Austria means a high quality of living and at the same time luxurious water conditions. A high percentage of its population is supplied with alpine spring water, which is known for superior quality and excellent taste. There are few countries in the world where everyone can drink tap water. Being advantageously situated is a good basis, but only half the truth. This well-functioning supply system in Austria is also due to a wise national management and not to forget also to strong transnational collaboration. Because of these to aspects of organization we can view Austria as an international role model for water management.

Being provided with this abundance of water, Austria statistically has a high water usage (compared to the global average) – summed up, the daily water use in an average household is 130 liters.

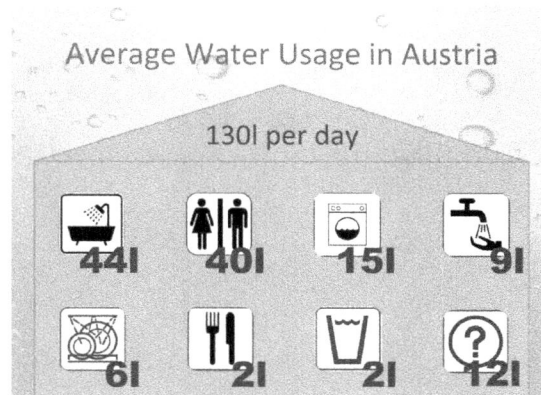

Figure 2: Average Water Usage in Austria

3 Conclusion

Now, after having examined some aspects of the legal basis of water management, ongoing national and international projects and sustainable water usage, we are much better informed about technical information, such as directives, settled laws and restrictions. Moreover we could enlarge our own awareness in terms of water usage. These new ways of perceiving water as an indispensable good made us change our habits. As we talked to our classmates about our insights, they reconsidered their habits too and consequently their awareness rose. Here from we draw the conclusion that education plays an important role concerning water management and usage, possibly even the very key role.

Acknowledgements

We would like to express our greatest gratitude to the people who have helped and supported us throughout the project.

First of all we want to thank our headmaster Dir. Edwin Scheiber, who gave us students the opportunity to participate at such a project.

Secondly we are grateful to our teacher Mr. Martin Windischhofer for his continuous support by his encouragement through ongoing advice and help to this day and for his decision to supervise us in Singapore.

A special thank of ours goes to Mrs. Angela Ransdorf, who once has been to Singapore and exchanged her interesting ideas and thoughts.

Finally we wish to thank our school sponsor "Helvetia", who took over the financing of this project.

References

[1]   http://duz.lebensministerium.at/duz/duz/category/76
       2111

[2]   http://www.lebensministerium.at/wasser/wasser-
       oesterreich.html

[3]   http://www.lebensministerium.at/en/fields/water/Water-and-the-public/Buildawareness.html

[4]   http://www.lebensministerium.at/wasser/wasser-oesterreich/plan_gewaesser_ngp/umsetzung_wasserrahmenrichtlinie/feg_fge.html

Fig. 1 http://commons.wikimedia.org/wiki/File:Autriche_hydro-de.svg

Fig. 2 Felix Oblin: Microsoft PowerPoint 2011

Water is power

Anna Todsen, Cecilie Stjernholm, Jacob Schouw and Christian Trangbæk

Eisbjerghus Efterskole, Nørre Aaby, Denmark

Abstract

Based on a historical approach to water as a geopolitical resource, this study have investigated how control over water resources has given power, and how that power has impacted on local, regional and global conflicts. Using the results, the study have also looked at what might happen in the future. Many of the world's conflicts take place in areas of the world where there are, or could be, water shortages, so it is highly unlikely that those wider conflicts can be resolved if the water conflicts are not solved at the same time. This study do not provide "solutions", but highlight what needs to be done if the water issues are to be resolved.

Keywords

Power, conflict, control

Introduction

Water is life; without it, humans will die of thirst or starve to death. However, water is more than just life, it represents power. Throughout history, it can be shown that those who control and master water resources also obtain power; power over other societies, power in the form of energy, economic power. Conversely, many of the conflicts that have taken place in the world result from disputes or conflicts over water resources, conflicts over that power.

The German philosopher Georg Hegel once famously said that "we learn from history that we do not learn from history". If we want the world to be a better place in the 21st century, perhaps we should learn from what has taken place in the past, and recognize that one of the keys to future peace and prosperity is better management of water resources. Many of the world's potential flashpoints are in areas of water stress; if we want to do something about the former, then we need to do something about the latter.

Investigation and Discussion

Ancient Egypt, An Example of Offensive Power

How can I rule the Middle East? That's a question many kings in ancient times have asked themselves. The answer was; by controlling the Nile.

The Nile was created around 3,600 BC when the great rain stopped. What had once been fertile lands became barren desserts, and the only fertile areas were around the river itself. Its delta was filled with food. The Nile did not just support crops but also fish and exotic animals, which lived in the fertile lands. The Nile made possible an entire culture as never seen before. Before the year 3,000 BC the fertile lands around the Nile were not ruled by one man. But around the year 3,000 BC a man took control over the lands

with a large army, and founded The Old Kingdom. The Nile gave them food and a lot of trade opportunities. Huge fish such as the Nile Perch could feed many families. Later on the people around the Nile built the pyramids, with methods that we still cannot fully understand, but we do know that it's because of the Nile that they could transport the many tons of stones to the building sites.

The people prospered, they never starved, so there just became more and more people in Egypt, who could work on the large building sites, where the Pharaohs' visions were made a reality. They could also create the largest army in the region and hire mercenaries from others. When the Pharaohs ruled Egypt, they had over 20.000 men ready to defend the Nile, because many other people wanted it; Hittites, Libyans, Hyksos. They were surrounded by people that desired the Nile, and the riches that came with it.

Yet large armies were not just available for defense, they could also attack. Alexander the Great used Egypt as a stable food supply, so his armies in Asia never starved. The Roman Empire also needed the Nile, because they had to feed 1, 5 million people in Rome alone, and for free. One of the main reasons the Empire could keep on growing was that Egypt could keep it supplied.

But why was the Nile so important? There are two reasons.

1. It provided a stable food chamber. Every year, and at the same time every year, the Nile flooded the areas along its banks. And when it withdrew to its normal state, the silt left behind fertilized the flooded areas, leaving it ready for seeding. The resultant harvest was large enough to feed millions of people, as well as ensuring stable food supplies if you wanted to invade a neighboring country.

2. The large amount of water secured the population's survival and gave control over a large proportion of all the food generated in the Middle East. That gave Egypt's leaders a strategic and diplomatic advantage, because according to the teachings of Sun Tzu; "The man that controls the food, controls the people".

Control the Nile, and one could control the region.

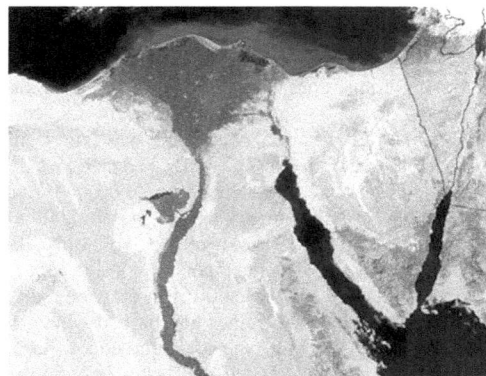

Figure 1: The Blue Nile Falls fed by Lake Tana near the city of Bahir Dar, Ethiopia.

181

17th Century Holland, An Example of Defensive Power

A large proportion of the land mass of the modern Netherlands lies below sea level. The Dutch have always striven to control the waters around them, and the first official water board was established as long ago as 1255.

However, draining the land did not just provide a greater area on which to grow crops. The many rivers running through the country, along with the fact that it's a very flat and low-lying gave the Dutch an opportunity, namely a flooding system which they could use to defend themselves against enemy invaders. The idea was first conceived at the siege of Leiden during the 80 year-long war of independence against the Spanish. The rebels managed to flood the surrounding area and sail in with the rebel fleet, thereby lifting the siege in October 1574.

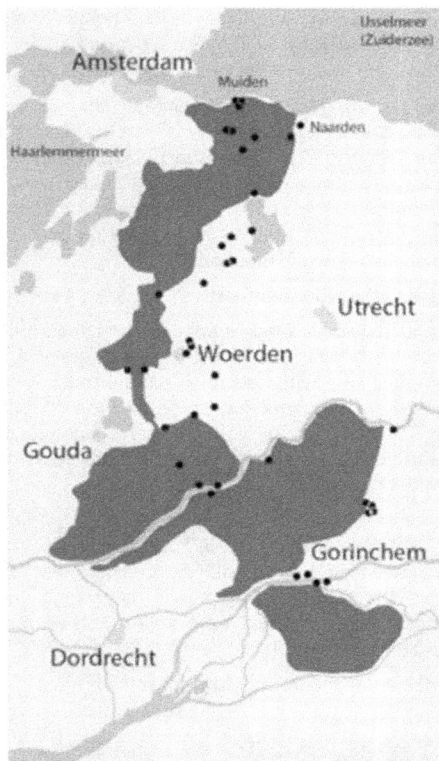

Figure 2: The old Hollandic Waterline

Flooding low-lying areas by diverting either river or sea water proved to be a very useful tactic in the case of defending. The old Hollandic water line was built between 1629 and 1815 by connecting different bodies of water and diverting the Waal River with sluices and dykes, effectively turning The Netherlands into an island. The water flow was carefully regulated in order to make the depth only 40 cm, so the use of ships by attackers was ruled out. The defenders would also use hidden traps in the water, such as spikes, deep holes and later mines and barbed wire. All the weak points along the line where reinforced by forts and other stationary defences. When the water froze in winter, the Dutch where able to manipulate the water levels so that the ice would be too thin to carry a man's weight, alternatively the ice could be broken up so as to create an

obstacle. In 1672 Louis XIV of France, the most powerful ruler in Europe, set out to conquer the United Provinces of the Netherlands. He and his army were stopped by the Hollandic water line and a peace treaty was signed.

The line had one great weakness, namely when there was a great freeze. In the terrible winter of 1794-1795 the French revolutionary army was able to overcome the line because it had frozen solid. When the United Kingdom of the Netherlands was formed after Napoleon's defeat at the Battle of Waterloo in 1815, the old Hollandic water line was modernized over the following 100 years. However, during its heyday during the 17th century, it protected a vulnerable new nation from being swallowed up by its bigger neighbours.

18th Century Britain, An Example of Economic Power

The development of canals was perhaps the most important factor behind the Industrial Revolution in England. Huge amounts of heavy goods such as coal, iron and bricks had to be moved, but the roads could not handle such weights and the vehicles needed did not exist. The canals were also needed for the transport of delicate products, such as glass and china that would break if transported on badly made, bumpy roads. The idea was that it should be possible to construct waterways to go where they were needed. This would allow goods to be transported quickly and cheaply.

The man most associated with early British canals was the Duke of Bridgewater. He was a coalmine owner in Lancashire and needed to get the coal transported the six miles to Manchester where the big market was. The duke hired the engineer James Brindley to design and build the canal. It took two years to complete, and had a series of tunnels that were linked directly to the coalmines.

Figure 3: The Bridgewater Canal

The canal was a huge success. The duke saved a lot of money on transportation, and the coal price fell in Manchester by 50%. The cheaper the price, the more coal was sold. Brindley also gained from the canal, he gained fame and more work.

Other people saw the success of the Bridgewater Canal and did likewise, opening up Britain even more with a series of canals that linked the major industrial centres of the country. Entrepreneurs invested vast sums of money in the canal projects, thereby stimulating the economy.

The heyday of canal building was in the period up to 1760, and again after the economic depression of the early 1780's; by the early 19th century, the canal network was complete. At that point, they were overtaken by another transport revolution, the invention and development of railways. But for about 50 years they provided the initial boost to economic development that propelled Britain to the forefront of world powers. By the late 19th century the small island country ruled about a quarter of the world, at the head of an empire on which the sun never set.

Modern India, An Example of Energy Power

A modern example of how water has been used as an energy source with economic advantages for poor regions is the Narmada River Valley project in India.

The Narmada River is the fifth longest river in India, with a total length of 1312 km. The river flows through Madhya Pradesh, Maharashtra and Gujarat before emptying into the Arabian Sea at Bharuch. It's called "the lifeline of Madhya Pradesh" because 87% of the river runs in Madhya Pradesh, making the soil fertile and so possible to cultivate agriculture. All in all, the river and its 41 tributaries drain about 98,796 sq. km of land.

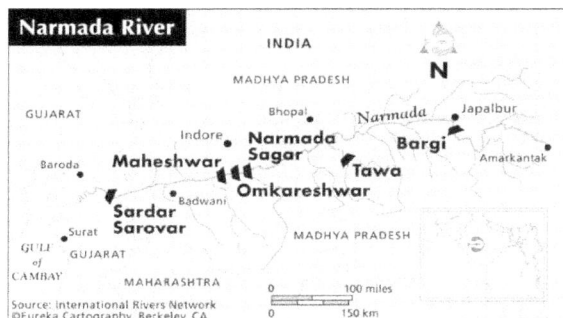

Figure 4: Flow of the Narmada River from the state of Madhya Pradesh through the state of Maharashtra and finally into Gujarat, emptying in the Arabian Sea at Bharuch.

The Narmada is an essential resource for a land with serious issues in maintaining a constant water supply throughout the year. Most of India is in the "monsoon climate zone", characterized by abundant rainfall in short periods. In Gujarat, Maharashtra and Madhya Pradesh the rainfall is concentrated into the four months between June and September. When it rains, Indians must resist the heavy rainfall and the danger of flooding. But for the rest of the year the inhabitants must draw their water from lakes, rivers and storage tanks and from groundwater sources.

The construction of hydro-electric dams on the Narmada river, and their impact on the millions of people dependent on its water supply, have become one of the most discussed and important social issues in modern India. Discussions started after independence in the 1940's, but were not realised until 1979. The delay was caused by the fact that each of the three states through which the Narmada River flows put forward its own scheme to harness its irrigation and power potential; they argued about where the dams should be built, who should build them, how the water made available for irrigation should be divided among them, and what kind of power plants should be constructed. In

particular, they argued about how high the main dam should be; Gujarat wanted a higher dam in order to maximize water supply; the other two states objected because it would lead to greater flooding. Finally, after decades of debate, the Indian Supreme Court accepted that the construction of the dams should start. Construction began in 1979, and the whole project is expected to be finished in 2025.

Many large dams have already been constructed, including the main one, the Sardar Sarovar Dam, the world's second largest. The main power plant consists of six 200MW pump turbines to generate electricity. The project will function as a newly created energy source in an area with high and unmet power demand, and so act as a spur to development in the involved regions.

Figure 5: The Sardar Sarovar Dam in Navagam, Gujarat.

Historical Summary

The above are historical examples of how the control of water resources has led to extra power. But they have also led to extra conflict. In the cases of Ancient Egypt and Holland, those conflicts were the direct result of the control of the water. In the case of Britain, canals did not lead directly to war. However, they did lead to greater economic prosperity, which allowed the building-up of a huge colonial empire. And when that empire eventually unraveled, there was a lot of war; during Partition in India, the Mau Mau rebellion in Kenya, the Communist insurgency in Malaya. While in India, although the economic benefits of the dam have been positive, there have also been conflicts, as villagers are displaced from the homes which have had to make way for the dams' reservoirs, or people object to the adverse effects on nature.

Looking forward in time, it is quite possible that similar conflicts will also be caused by, or connected to, water disputes. We have looked at four parts of the world, where problems may arise. All involve water, though the other factors are very different.

The Jordan Valley, Political Tension

The Israel-Palestine conflict has been raging for decades. It did not start because of water and is not continuing because of water, political factors are paramount. However water supply has become a powerful weapon in the conflict. But why? And what is being done about it?

The fact of the matter is that Israel is an occupying power, and has 100% control of the water supply in the West Bank and Gaza Strip. There is a huge gap in the water consumption per capita between Palestinians living in the West Bank and Gaza and Israelis living in Israel proper.

Furthermore, the quality of the water is substantially lower in the West Bank and Gaza compared with that in Israel. This results in diseases and pollutants ending up in Palestinian drinking water. Another problem is the fact that water is 8 times more expensive for a Palestinian than for an Israeli citizen and 20 times as expensive compared with that supplied to an Israeli settler. Consequently, this makes farming very expensive for Palestinians, especially in the dry summer months. Palestinians are very dependent on water since a large percentage of their GDP comes from agriculture while only 3 percent of Israel's GDP comes from agriculture.

Palestinians also need permission from Israeli authorities in order for them to dig wells. Only a handful of Palestinian wells have been authorized and hence dug.

It is clear that the Israeli government is using water as a weapon in its political fight for control of the land. The issue is not a lack of water per se, since the river Jordan and other water sources would suffice for everybody. However, because the available water is shared so dramatically in Israel's favour, most Palestinian villages lack adequate water facilities. They often have to buy water from privately owned Israeli companies in order to meet their demands. Hence we can conclude that water is being used as a weapon by the Israeli Government to control and weaken the Palestinian economy.

Nothing is being done from the Israeli side, and it is clear that they intend to preserve this efficient weapon and grip on the Palestinian economy. Possible political progress takes as its starting point the Oslo accords. It is relevant in this context that they do not mention any measures to improve the Palestinian water situation.

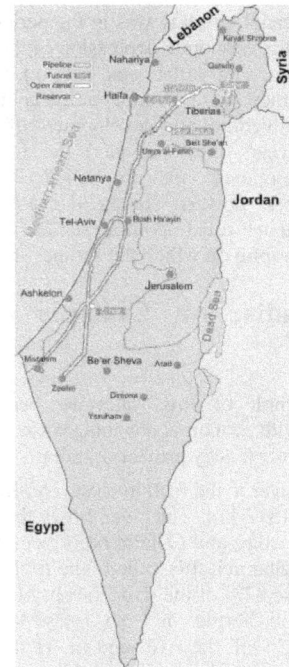

Figure 6

The Nile Basin, Upstream Population Growth

In 2100, it is expected that a number of countries in Africa will be in the top ten most populated countries in the world. Today, Egypt is the most populous country in the Nile River Basin; but in 2100, its population will be dwarfed by that of Ethiopia, Tanzania, Uganda and Kenya.

Today, these countries have enough water, and the water is shared by way of a riparian treaty, which allows each country to take a fixed amount of the water flowing down the river. The riparian treaty was drawn up when nearly all of the relevant countries formed part of the British Empire, and has worked reasonably well. However, as a greater share of future population growth takes place in the upstream countries, so the pressure will grow to change the shares enshrined in the treaty.

If Tanzania and Uganda started taking more water from Lake Victoria, thereby breaking the riparian treaty, it could cause many problems. Uganda and Tanzania would most likely deny that they should follow the treaty, because it was forced upon them by the British Empire. Egypt could in turn threaten to use military force in order to protect its water supplies; as the biggest power in the region, it would probably win.

In this way, a conflict about water might easily lead to a wider conflict. How could this be prevented? It is clear that the treaty should be renegotiated and updated, so they can avoid conflict; but it will not be an easy task.

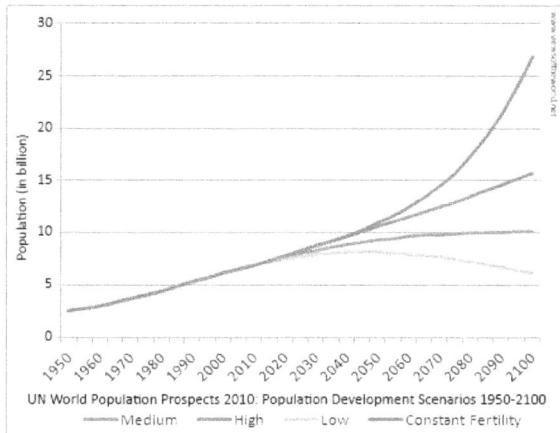

Figure 7

Central Asia and the Aral Sea, Pollution and Environmental Degradation

The USSR was the world's largest country by area, stretching from the Baltic and Black Seas to the Pacific Ocean. Founded in 1922 under communist rule, it established an economic system based on state ownership, collective farming, industrial manufacturing and the distribution of goods centralized and directed by the government. Priority was given to heavy industry with little or no attention paid to environmental effects.

Only after the collapse of the Soviet Union in late 1991 did the true scale of this environmental crisis become apparent. The water supplies of most major cities are undrinkable, beaches are frequently closed because of pollution, rivers in the European part of the country are off limits because they are so filthy, and fields have been polluted because of the heavy use of pesticides and fertilizers.

One of the most famous cases of pollution involves the Aral Sea, located in central Asia between what is now Kazakhstan and Uzbekistan, in an area with deserts. The Aral Sea was once the fourth largest saline lake, with the main volume of water coming from high glaciers feeding into two main rivers, the Amu Darya and Dyr Darya, which emptied into the sea. In ancient times, the Sea was an oasis, where thousands of fishermen, farmers, merchants, hunters and craftsmen did well. I was once an important area that connected Europe and Asia as a part of the Great Silk Road. It supported a population of more than 60 million people with drinking water, and a successful fishing industry, which played a fundamental role in the economy and wellbeing of the region.

Figure 8: The Aral Sea is located between now Kazakhstan and Uzbekistan in Central Asia

Soon after the creation of the Soviet Union, its leaders made plans to increase the production of cotton in Central Asia by expanding irrigation. So in the 1960's, the Soviet government decided to divert the rivers feeding the Aral Sea, so that they could irrigate the deserts surrounding the Sea, instead of running into the sea itself. But their irrigation techniques were inefficient, and caused a lot of waste. This turned out to have grave consequences for the region. It quickly became visible that the Aral Sea was drying out, and by the 1980's, during dry or average years, no river water reached the Sea. By then, the Aral Sea's surface area has shrunk by approximately 74% and its total volume by almost 85%. The Sea now receives only a tenth of the water that it used to do.

The changes in the area and volume of the Aral Sea have had a serious impact on the environment, livelihood and economies of local populations in Central Asia. There have been various examples of unexpected climate feedbacks and public health issues, affecting the lives of millions of people in the region. The region is now heavily polluted and people are suffering from a lack of drinking water, the inability to cultivate crops and by lung diseases. The withdrawn sea has left huge areas covered with salt and toxic chemicals, which are carried away by the wind as a kind of toxic dust, spreading to surrounding areas. The population around the Aral Sea now shows high rates of lung cancer and other diseases. The salinization of the land is also making agriculture in the region impossible, and the fishing industry is destroyed. The decline of the Aral Sea has been associated with the loss of thousands of jobs.

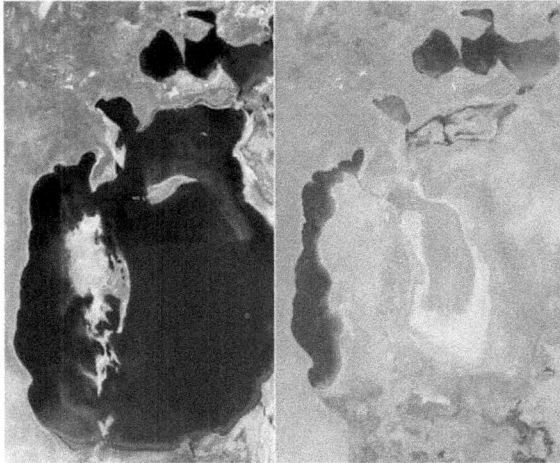

Figure 9: Changes seen in the Aral Sea between 1989 (left) to 2008 (right)

The United Nations has estimated that what remains of the sea will essentially disappear by 2020 if nothing is done to reverse its decline. However, whereas in the past the water basin was wholly within one country, the Soviet Union, now it is split between Tajikistan, Kyrgyzstan, Kazakhstan, Turkmenistan and Uzbekistan. Tensions between the five new countries over how to deal with the water and associated pollution problems could easily lead to a wider conflict, if one of them pursued a policy detrimental to the others.

An example is Kazakhstan's recent unilateral decision to build a dam. In the last decade, the Aral Sea has separated into two – the small Aral Sea fed by the Syr Darya and the Large Aral Sea fed by the Amu Darya. In a last-ditch attempt to save some of the remaining Sea, Kazakhstan built a dam between the northern and southern parts. This has led to progress in the northern sea, but has also had the unfortunate consequence of being a death sentence for the southern part, which was judged to be beyond saving. What was good for Kazakhstan was not necessarily good for the others.

The Himalayan Glaciers, Global Warming and Water Measurement

World electricity consumption is expected to grow by more than 56% during the next 30 years but we are also aware that global warming is an increasing problem. Renewable energy sources are the way forward and hydropower is an obvious option for the areas adjacent to the glaciers in the Himalayas.

Running 2000 kilometers from east to west and comprising more than 60000 km2 of ice, the Hindu Kush–Karakoram–Himalayan glaciers are a source of water for the quarter of the global population that lives in south Asia. Glaciers are natural stores and regulators of water supply to rivers, which in turn provide water for domestic and industrial consumption, energy generation, and irrigation.

Because of global warming, Himalayan glaciers lost an estimated amount of 174 gigatonnes of water between 2003 and 2009. That is a serious amount - melt water from the glaciers provides the power for a huge number of hydroelectric dams on the rivers near the Himalayan Mountains.

Hydroelectric dams are right now facing many difficulties mainly considering climatic and politico-economic factors. An example of this can be seen on the Brahmaputra, one of the largest rivers in the world. It rises in Tibet, travels through the mountains to India, and then heads to Bangladesh. India and China are building dams on the river, causing stir in the population. China and India have been accused of starting a "race to dam" the Brahmaputra. Both governments claim that the flow will not be affected and that all possible impacts have been considered. Despite the statements, there is no official water-sharing deal between India and China - just an agreement to share monsoon flood data.

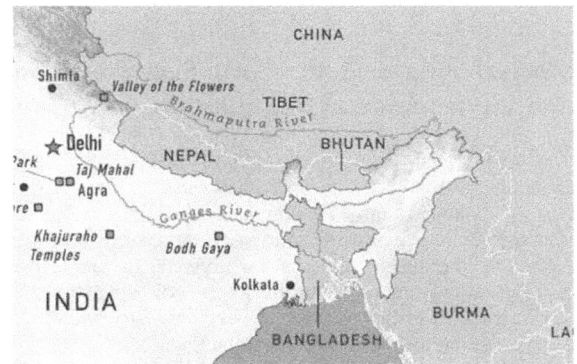

Figure 10: Brahmaputra River, one of the largest rivers in the world.

Hydroelectric power must play a role in South Asia's future, but in order for it to work, governments surrounding the Himalayas have to work together to measure the melt water, changing the river flows and constructing dams. Political obstacles must be overcome; collaboration between national borders and dialogue might be the answer. Unilateral action could adversely affect others, particularly Bangladesh. That in turn could lead to conflict.

Conclusion

History has shown that control of water has led to more power and more conflict. With more people, yet essentially the same amount of water in the world, it is quite possible that water has the potential to create even more conflict. We have identified four potential flashpoints, in which water will play a key role. There will be others.

Solving the water issues will not necessarily solve the conflicts; other factors might end up being more important. However, none of the potential conflicts will be truly solved unless and until the water issues are solved.

Water is life; water is power.

References

[1] http://www.historylearningsite.co.uk/canals_1750_to_1900.htm

[2] http://www.saburchill.com/history/chapters/IR/022.html

[3] http://www.mikeclarke.myzen.co.uk/Englishcanals.htm

[4] Illustration available at:
http://www.umich.edu/~snre492/Jones/narmada.html

[5] Picture available at:
http://www.panoramio.com/photo/28421979

[6] http://www.narmada.org/nvdp.dams/index.html

[7] http://en.wikipedia.org/wiki/Narmada_Valley_Development_Authority

[8] http://en.wikipedia.org/wiki/Narmada_River

[9] http://scholarworks.umass.edu/cgi/viewcontent.cgi?article=1015&context=edethicsinscience

[10] http://weather.about.com/od/monsoons/f/monsoons.htm

[11] http://www.ecoindia.com/rivers/narmada.html

[12] http://scholarworks.umass.edu/cgi/viewcontent.cgi?article=1015&context=edethicsinscience

[13] http://www.narmada.org/sardar-sarovar/sc.ruling/majority.judgement.htm

[14] http://www.umich.edu/~snre492/Jones/narmada.html#Background

[15] http://scholarworks.umass.edu/cgi/viewcontent.cgi?article=1015&context=edethicsinscience

[16] http://scholarworks.umass.edu/cgi/viewcontent.cgi?article=1015&context=edethicsinscience

[17] http://scholarworks.umass.edu/cgi/viewcontent.cgi?article=1015&context=edethicsinscience

[18] http://www.narmada.org/gcg/gcg.html

[19] http://www.narmada.org/nvdp.dams/1.

[20] http://www.bbc.com/news/magazine-26512465

[21] http://www.bbc.com/news/world-asia-india-26663820

[22] http://www.nature.com/news/climate-change-melting-glaciers-bring-energy-uncertainty-1.14031

www.ingramcontent.com/pod-product-compliance
Lightning Source LLC
Chambersburg PA
CBHW062018210326
41458CB00075B/6208